Big Data and Visual Analytics

Sang C. Suh • Thomas Anthony
Editors

Big Data and Visual Analytics

Editors
Sang C. Suh
Department of Computer Science
Texas A&M University-Commerce
Commerce, TX, USA

Thomas Anthony
Department of Electrical and
Computer Engineering
The University of Alabama at Birmingham
Birmingham, AL, USA

ISBN 978-3-319-87670-2 ISBN 978-3-319-63917-8 (eBook)
https://doi.org/10.1007/978-3-319-63917-8

Softcover re-print of the Hardcover 1st edition 2017

Printed on acid-free paper

This Springer imprint is published by Springer Nature
The registered company is Springer International Publishing AG
The registered company address is: Gewerbestrasse 11, 6330 Cham, Switzerland

Foreword

The editors of this book, an accomplished senior data scientist and systems engineer, Thomas Anthony, and an academic leader, Dr. Sang Suh, with broad expertise from artificial intelligence to data analytics, constitute a perfect team to achieve the goal of compiling a book on *Big Data and Visual Analytics*. For most uninitiated professionals "Big Data" is nothing but a buzzword or a new fad. For the people in the trenches, such as Sang and Thomas, Big Data and associated analytics is a matter of serious business. I am honored to have been given the opportunity to review their compiled volume and write a foreword to it.

After reviewing the chapters, I realized that they have developed a comprehensive book for data scientists and students by taking into account both theoretical and practical aspects of this critical and growing area of interest. The presentations are broad and deep as the need arise. In addition to covering all critical processes involving data science, they have uniquely provided very practical visual analytics applications so that the reader learns from the perspective executed as an engineering discipline. This style of presentation is a unique contribution to this new and growing area and places this book at the top of the list of comparable books.

The chapters covered are 1. Automated Detection of Central Retinal Vein Occlusion Using Convolutional Neural Network, by "Bismita Choudhury, Patrick H. H. Then, and Valliappan Raman"; 2. Swarm Intelligence Applied to Big Data Analytics for Rescue Operations with RASEN Sensor Networks, by "U. John Tanik, Yuehua Wang, and Serkan G. ldal"; 3. Gender Classification Based on Deep Learning, by "Dhiraj Gharana, Sang Suh, and Mingon Kang"; 4. Social and Organizational Culture in Korea and Women's Career Development, by "Choonhee Yang and Yongman Kwon"; 5. Big Data Framework for Agile Business (BDFAB) as a Basis for Developing Holistic Strategies in Big Data Adoption, by "Bhuvan Unhelkar"; 6. Scalable Gene Sequence Analysis on Spark, by "Muthahar Syed, Jinoh Kim, and Taehyun Hwang"; 7. Big Sensor Data Acquisition and Archiving with Compression, by "Dongeun Lee"; 8. Advanced High Performance Computing for Big Data Local Visual Meaning, "Ozgur Aksu"; 9. Transdisciplinary Benefits of Convergence in Big Data Analytics, "U. John Tanik and Darrell Fielder"; 10. A Big Data Analytics Approach in Medical Image Segmentation Using Deep Convolutional

Neural Networks, by "Zheng Zhang, David Odaibo, and Murat M. Tanik"; 11. Big Data in Libraries, by "Robert Olendorf and Yan Wang"; 12. A Framework for Social Network Sentiment Analysis Using Big Data Analytics, by "Bharat Sri Harsha Karpurapu and Leon Jololian"; 13. Big Data Analytics and Visualization: Finance, by "P. Shyam and Larry Mave"; 14. Study of Hardware Trojans in a Closed Loop Control System for an Internet-of-Things Application, by "Ranveer Kumar and Karthikeyan Lingasubramanian"; 15. High Performance/Throughput Computing Workflow for a Neuro-Imaging Application: Provenance and Approaches, by "T. Anthony, J. P. Robinson, J. Marstrander, G. Brook, M. Horton, and F. Skidmore."

The review of the above diverse content convinces me that the promise of the wide application of big data becomes abundantly evident. A comprehensive transdisciplinary approach is also evident from the list of chapters. At this point I have to invoke the roadmap published by the National Academy of Sciences titled "Convergence: Facilitating Transdisciplinary Integration of Life Sciences, Physical Sciences, Engineering, and Beyond" (ISBN 978-0-309-30151-0). This document and its NSF counterpart states convergence as "an approach to problem solving that cuts across disciplinary boundaries. It integrates knowledge, tools, and ways of thinking from life and health sciences, physical, mathematical, and computational sciences, engineering disciplines, and beyond to form a comprehensive synthetic framework for tackling scientific and societal challenges that exist at the interfaces of multiple fields." Big data and associated analytics is a twenty-first century area of interest, providing a transdisciplinary framework to the problems that can be addressed with convergence.

Interestingly, the Society for Design and Process Science (SDPS), www.sdpsnet.org, which one of the authors has been involved with from the beginning, has been investigating convergence issues since 1995. The founding technical principle of SDPS has been to identify the unique "approach to problem solving that cuts across disciplinary boundaries." The answer was the observation that the notions of Design and Process cut across all disciplines and they should be studied scientifically in their own merits, while being applied for developing the engineering of artifacts. This book brings the design and process matters to the forefront through the study of data science and, as such, brings an important perspective on convergence. Incidentally, the SDPS 2017 conference was dedicated to "Convergence Solutions."

SDPS is an international, cross-disciplinary, multicultural organization dedicated to transformative research and education through transdisciplinary means. SDPS celebrated its twenty-second year during the SDPS 2017 conference with emphasis on convergence. Civilizations depend on technology and technology comes from knowledge. The integration of knowledge is the key for the twenty-first century problems. Data science in general and Big Data Visual Analytics in particular are part of the answer to our growing problems.

This book is a timely addition to serve data science and visual analytics community of students and scientists. We hope that it will be published on time to be distributed during the SDPS 2018 conference. The comprehensive and practical nature of the book, addressing complex twenty-first century engineering problems in a transdisciplinary manner, is something to be celebrated. I am, as one of the founders of SDPS, a military and commercial systems developer, industrial grade software developer, and a teacher, very honored to write this foreword for this important practical book. I am convinced that it will take its rightful place in this growing area of importance.

Electrical and Computer Engineering Department
UAB, Birmingham, AL, USA

Murat M. Tanik
Wallace R. Bunn
Endowed Professor of
Telecommunications

Contents

Automated Detection of Central Retinal Vein Occlusion Using Convolutional Neural Network

Bismita Choudhury, Patrick H.H. Then, and Valliappan Raman

Abstract The Central Retinal Vein Occlusion (CRVO) is the next supreme reason for the vision loss among the elderly people, after the Diabetic Retinopathy. The CRVO causes abrupt, painless vision loss in the eye that can lead to visual impairment over the time. Therefore, the early diagnosis of CRVO is very important to prevent the complications related to CRVO. But, the early symptoms of CRVO are so subtle that manually observing those signs in the retina image by the ophthalmologists is difficult and time consuming process. There are automatic detection systems for diagnosing ocular disease, but their performance depends on various factors. The haemorrhages, the early sign of CRVO, can be of different size, color and texture from dot haemorrhages to flame shaped. For reliable detection of the haemorrhages of all types; multifaceted pattern recognition techniques are required. To analyse the tortuosity and dilation of the veins, complex mathematical analysis is required in order to extract those features. Moreover, the performance of such feature extraction methods and automatic detection system depends on the quality of the acquired image. In this chapter, we have proposed a prototype for automated detection of the CRVO using the deep learning approach. We have designed a Convolutional Neural Network (CNN) to recognize the retina with CRVO. The advantage of using CNN is that no extra feature extraction step is required. We have trained the CNN to learn the features from the retina images having CRVO and classify them from the normal retina image. We have obtained an accuracy of 97.56% for the recognition of CRVO.

Keywords Retinal vein occlusion • Central retinal vein occlusion • Convolution • Features

B. Choudhury (✉) • P.H.H. Then • V. Raman
Centre for Digital Futures and Faculty of Engineering, Computing and Science, Swinburne University of Technology, Sarawak Campus, Kuching, Sarawak, Malaysia
e-mail: bismi.choudhury@gmail.com

S.C. Suh, T. Anthony (eds.), *Big Data and Visual Analytics*,
https://doi.org/10.1007/978-3-319-63917-8_1

1 Introduction

The retina is the light sensitive tissue covering the interior surface of the eye. The cornea and the lens focus light rays on the retina. Then, the retina transforms the light received into the electrical impulses and sends to the brain via the optic nerve. Thereby, a person interprets those impulses as images. The cornea and the lens in the eye behave like the camera lens, while the retina is analogous to the film. Figure 1 shows the retina image and its different features like optic disc, fovea, macula and blood vessels.

The Retinal Vein Occlusion (RVO) is an obstruction of the small blood carrying veins those drain out the blood from the retina. There are one major artery, called the central retinal artery, and one major vein, called the central retinal vein, in the retina. The Central Retinal Vein Occlusion (CRVO) occurs when a thrombosis is formed in this vein and causes leaking of blood and excess fluid into the retina. This fluid often accumulates around the macula, the region for the central vision, in the retina. Sometimes the blockage occurs when the veins in the eye are too narrow [1].

The diagnostic criteria for CRVO are characterized by flame-shaped, dot or punctate retinal hemorrhages or both in all four quadrants of the retina, dilated and tortuous retinal veins, and optic disc swelling [2].

The CRVO can be either ischemic or non-ischemic. About 75% of the cases, non-ischemic CRVO is a less severe form of CRVO and usually has a chance for better visual acuity. Ischemic CRVO is a very severe stage of CRVO where significant complications arise and can lead to the vision loss and probably damage the eye [1].

Macular edema is the prime reason for the vision loss in CRVO. The fluid accumulated in the macular area of the retina causes swelling or edema of the macula. It causes the central vision of a person to become blurry. The patients

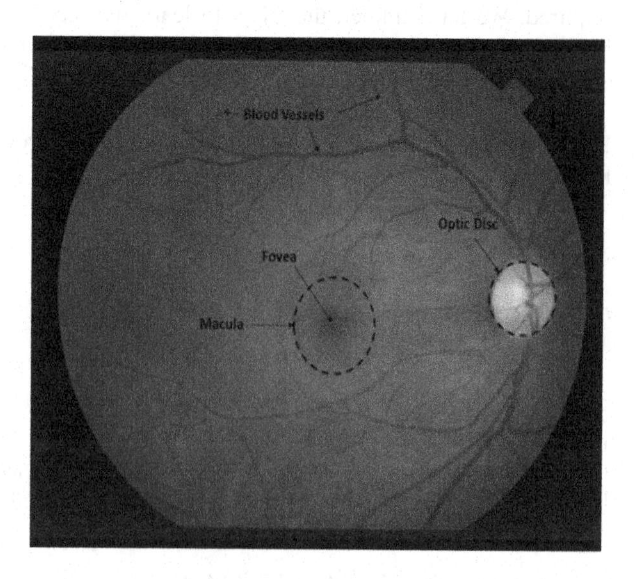

Fig. 1 Retina and its different features

with the macular edema following CRVO might have some of the most common symptoms, such as blurred vision, distorted vision, or vision loss in all or part of the eye [2].

The lack of oxygen (ischemia) in the retina can lead to the growth of the abnormal blood vessels. The patients with ischemic CRVO develop neovascular glaucoma over three months or longer period of time. In neovascular glaucoma, the abnormal blood vessels increase the pressure in the eye that can cause severe pain and vision loss [3].

Usually, the people who are aged 50 and older have higher chance of suffering from CRVO. The probability of occurring CRVO is higher in people having diabetes, high blood pressure, high cholesterol, or other health issues that interrupt blood flow. The symptoms of retinal vein occlusion can be range from indistinct to very distinct. Most of the time, just one eye suffers from painless blurring or loss of vision. Initially, the blurring or the vision loss of the eye might be minor, but this situation gets worse over the next few hours or days. Sometimes the patients might lose the complete vision almost immediately. In 6–17% of the cases, the second eye also develops the vein occlusion after the first one. Up to 34% of eyes with non-ischemic CRVO convert to ischemic CRVO over 3 years [1].

It is crucial to recognize CRVO to prevent further damage in the eye due to vein occlusion and treat all the possible risk factors to minimize the risk of the other eye to form CRVO. The risk factor of CRVO includes hypertension, diabetes, hyperlimidemia, blood hyperviscosity, vascular cerebral stroke and thrombophilia. The treatment of any of these risk factors reduces the risk of a further vein occlusion occurring in either eye. It may also help to reduce the risk of another blood vessel blockage, such as may happen in a stroke (affecting the brain) or a heart attack or, in those with rare blood disorders, a blocked vein in the leg (deep vein thrombosis) or lung (pulmonary embolism). There is no cure, but early treatment may improve vision or keep the vision from worsening [4].

The automatic detection of CRVO in the early stage can prevent the total vision loss. The automatic detection can save lots of time for the ophthalmologist. Rather than putting lots of effort in diagnosis, they can put more time and effort for the treatment. Thereby, the patients will receive the treatment as early as possible. It will be also beneficial for the hospitals and the patients in terms of saving time and money. For the diagnosis of retinal disease, mostly fluorescein angiographic image or color fundus image is taken. However, compared to angiographic images color fundus images are more popular in the literature of automatic retina analysis. Color fundus images are widely used because it is inexpensive, non-invasive, can store for future reference and ophthalmologists can examine those images in real time irrespective of time and location. In all the computer aided detection system, the abnormal lesions or features are detected to recognize the particular disease.

The early symptoms of CRVO are very subtle to detect. When non-ischemic CRVO forms, the retina remains moderately normal. So, there is higher chance that general automatic detection system fails to detect CRVO in the earliest stage.

Another clinical sign for CRVO is dilated tortuous vein. So, it is important to segment and analyse the retinal vasculature. Most importantly, it is required to detect the vein and calculate the tortuosity index and analyse the change in blood vessels due to dilation. Moreover, it is crucial to detect any newly generated blood vessels leading to neovascularization for ischemic CRVO. Another clinical characteristic of CRVO is haemorrhages, which can be of different size, color and texture. The haemorrhages in CRVO are mostly dot haemorrhage and flame shaped haemorrhage. The dot haemorrhages appear similar to the microaneurysms. Therefore, the segmentation process can't distinguish between the dot haemorrhage and the microaneurysm. Therefore, by default the dot haemorrhages are detected as microaneurysm by the automated microaneurysm detection process. The literature does not support much description about the automated detection of the haemorrhages [5]. In the ischemic stage, there will be multiple Cotton wool spots and the literature doesn't provide much attention to the automatic detection of cotton wool spots. In short, the problem with the automatic detection of the CRVO is that, the sophisticated segmentation and feature extraction algorithms are required for each of the clinical signs. For example, for reliable detection of the haemorrhages of all types; multifaceted pattern recognition techniques are required. For analysing dilated veins, tortuous vein, newly formed blood vessels, we need complicated mathematical approach for such feature extraction. Again, the performance of such algorithms and the classification depends on the image quality of the retina image acquired. The inter-and intra-image contrast, luminosity and color variability present in the images make it challenging to detect these abnormal features. To our best knowledge, no research work related to automatic detection of CRVO has been done yet.

In this chapter, we approached the deep learning method for detecting the CRVO. We have exploited the architecture of Convolutional Neural Network (CNN) and designed a new network to recognize CRVO. The advantage of using CNN is that, design of complex, sophisticated feature extraction algorithms for all the clinical signs of CRVO are not necessary. The convolution layer in the neural network extracts the features by itself. Moreover, CNN takes care of image size, quality etc. The chapter is organized as follows: the first part will briefly describe about the types of CRVO. In the second section, the computer aided detection system for medical images will be discussed. In the third section, we will review previous related work on the automated detection of vein occlusion. The fourth section will describe the theory and architecture of the Convolutional Neural Network. The fifth section will describe the design of the CNN for the recognition of CRVO.

2 Central Retinal Vein Occlusion (CRVO)

The two types of CRVO, ischemic and non-ischemic, have very different diagnoses and management criteria from each other. Both the types are briefly discussed below:

2.1 Non-Ischemic CRVO

It was reported that majority of the cases (about 70%), the patients suffer from non-ischemic CRVO [3]. Over 3 years, 34% of non-ischemic CRVO eyes progressed to ischemic CRVO. There is low risk of neovascularization in case of non-ischemic CRVO. The clinical features of non-ischemic CRVO are as follows:

- Vision acuity >20/200.
- The low risk of forming neovascularization.
- More dot & blot hemorrhages.
- The retina in non-ischemic CRVO will be moderately normal.
- There is no Afferent Pupillary Defect (APD).

Figure 2 shows the retina image with non-ischemic CRVO.

2.2 Ischemic CRVO

According to the fluorescein angiographic evidence, the ischemic CRVO is defined as of more than 10 disc areas of capillary non-perfusion on seven-field fundus fluorescein angiography [1]. It is associated with an increased risk of neovascularization and has a worse prognosis [3, 6]. There is a 30% chance of converting non-ischemic to ischemic CRVO [1]. More than 90% of patients with ischemic CRVO have a final visual acuity of 6/60 or worse [6]. The clinical features of ischemic CRVO are as follows:

- Visual acuity <20/200
- The high risk of forming neovascularization
- Widespread superficial hemorrhages.

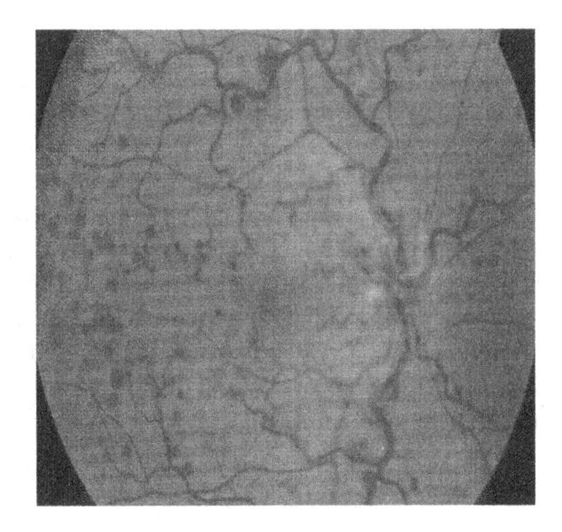

Fig. 2 Non-ischemic CRVO

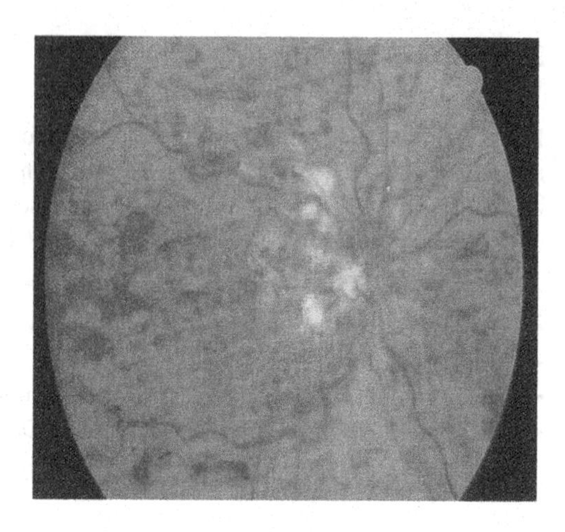

Fig. 3 Ischemic CRVO

- Multiple Cotton Wool Spots.
- Poor capillary perfusion (ten or more cotton wool spots or ten DD capillary non-perfusion on fluorescein angiography).
- Turbid, orange, edematous retina.
- Poor prognosis
- Degree of retinal vein dilatation and tortuosity.
- High Relative Afferent Pupillary Defect (+RAPD).

Figure 3 shows the retina image with ischemic CRVO.

3 Computer Aided Detection (CAD)

The Computer Aided Detection (CAD) systems are designed to assist physicians in the evaluation of medical images. CAD is rapidly growing in the field of radiology to improve the accuracy and consistency of the radiologists' image interpretation. CAD system processes digital images and highlight the suspicious section to evaluate the possible disease. The goal of the CAD systems is to detect the earliest signs of abnormality in the patients' medical image that human professionals cannot. It is pattern recognition software that automatically detects the suspicious features in the image to get the attention from the radiologist and reduce the false negative reading. The computer algorithm for automatic detection usually consists of multiple steps, including image processing, image feature extraction, and data classification through a different classifier such as artificial neural networks (ANN) [7]. The CAD systems are being used for different image modalities from Magnetic Resonance Imaging (MRI), Computed Tomography (CT), ultrasound imaging, retinal funduscopic image etc.

similarities of neighboring pixels within the fundus image. In the second method, Gabor functions are used as template matching procedure to calculate the response. The fractal analysis was performed on a number of retinal images by combining the segmented images and their skeletonized versions. In fractal analysis, the box-dimension is used to estimate the fractal dimension via box counting. The lower and upper box counting dimensions of a subset, respectively, are defined as follows:

$$\underline{dim}_B(F) = \frac{\lim}{r \to 0} \frac{\log N_r(F)}{-\log r}; \overline{\dim}_B(F) = \frac{\overline{\lim}}{r \to 0} \frac{\log N_r(F)}{-\log r} \quad (1)$$

If the lower and upper box-counting dimensions are equal, then their common value is referred to as the box-counting dimension of F and is denoted with

$$dim_B(F) = \lim_{r \to 0} \frac{\log N_r(F)}{-\log r} \quad (2)$$

where $N_r(F)$ can be of the following:" (1) the smallest number of closed balls (i.e., disks, spheres) of radius $r > 0$ that cover F; (2) the smallest number of cubes of side r that cover F; (3) the number of r -mesh cubes that intersect F; (4) the smallest number of sets of diameter at most r that cover F; (5) the largest number of disjoint balls of radius r with centers in F."

The fractal dimension is calculated for both the skeletonized images of normal retina and the retina with RVO. There is no significant difference in the fractal dimension of healthy eyes. But, the fractal dimension is quite visible in case of retina with CRVO. The fractal dimensions computed seemed to be beneficial in separating the different types of RVO.

4.3 Deep Learning Approach

Zhao et al. in [11] proposed a patch based and an image based voting method for the recognition of BRVO. They exploited Convolutional Neural Network (CNN) to classify the normal and BRVO color fundus images. They extracted the green channel of the color fundus image and performed image preprocessing to improve the image quality. In the patch based method they divided the whole image into small patches and put labels on each patches to train the CNN. If the patch has BRVO features, labeled as BRVO otherwise labeled as normal. During the training phase, only the patches with the obvious BRVO feature are labeled as BRVO. Those ambiguous patches are discarded. The testing is done by feeding all the patches of a test image to the trained CNN. They kept the threshold of 15 patches for each test. If the test image passes the threshold, the testing image is classified to BRVO. In the image based scheme, at first, three operations, noise adding, flipping, and rotation are performed on a preprocessed image. Depending on the classification results of these four images, the final decision for a test image is made. "If the classification

Table 1 Summary of the research works done for identifying RVO

Author	Year	Type of RVO	Method	Accuracy
Zhang et al. [8]	2015	BRVO	Hierarchical linear binary pattern, support vector machine	96.1%
Gayathri et al. [9]	2014	–	Complete linear binary pattern, neural network	–
Fazekas et al. [10]	2015	CRVO, BRVO, HRVO	Fractal properties of blood vessels	–
Zhao et al. [11]	2014	BRVO	Convolutional neural network	97%

results of the three new images (noisy, flipped, and rotated) are the same, the test image is classified in the class of the three new images. Otherwise, the test image is classified to the class of the original image". For patch based method they obtained 98.5% and for image based method they obtained 97%. Compared to patch based method, image based voting method is more practical.

The Table 1 summarize the various research done for the automated diagnosis of retinal vein occlusion

From the limited previous research work, it is clear that less attention has been paid towards the automatic detection of RVO. The fractal analysis described in [10] provided the calculation of fractal dimension of RVO images and the possibility of using those values for quantifying the different types of RVO. No clear information is provided regarding the accuracy of the methods described in [9, 10]. In [9, 10], retinal vascular structure of the color fundus image is analysed to extract the features and used classifier to detect RVO. In [8], the features are extracted from the whole image to detect BRVO in Fluorescein Angiography (FA) image. In [11], CNN is used to detect the BRVO in color fundus image. The majority of the available research works are on automatic detection of BRVO. No research work has been found that focuses on detecting haemorrhages, analyse vessel tortuosity and dilation to recognize RVO. There is no existing method for the automatic detection of CRVO considering the fact that the visual imparity is more severe in case of CRVO. Hence, it is very important to design an automatic detection system for recognizing CRVO. For automatic detection of CRVO there can be two approaches. One, individually extract the abnormal features from the segmented retinal pathologies by compound pattern recognition techniques and then, fed them to a classifier to identify the CRVO. Otherwise, extract the abnormal features from the whole image and use supervised machine learning classifier to identify CRVO.

5 The Proposed Methodology

We opted a deep learning approach to extract the features from the raw retina image and classify as CRVO. We explored the architecture of the Convolutional Neural Network (CNN) and designed a CNN for learning the CRVO features and classify the CRVO images from the normal retina images. There are some advantages of using CNN. First, we do not have to design individual segmentation and feature extraction algorithms for all the clinical signs of CRVO (haemorrhages, cotton wool spots, dilated tortuous veins and newly formed blood vessels) and rely on the accuracy of such algorithms to classify CRVO. Second, the convolutional neural network is invariant to any kind of distortion in the image, for example, the different lighting condition, camera position, partial occlusion etc. Third, easier to train compared to conventional Neural Network (NN) due to reduced parameters used during training. Fourth, memory requirement is less as convolution layer use same parameters to extract the features across the different locations in an image. In this study, we collected the normal retina image and retina with CRVO image from multiple publicly available databases. We used STARE database (http://cecas.clemson.edu/~ahoover/stare/), DRIVE database (http://www.isi.uu.nl/Research/Databases/DRIVE/), dataset of Dr. Hossein Rabbani (https://sites.google.com/site/hosseinrabbanikhorasgani/datasets-1) and Retina Image Bank (http://imagebank.asrs.org/discover-new/files/1/25?q). The existing methods conducted their experiments on different datasets and since those datasets have different image size and quality we cannot compare their performance directly. Because, the experimental results on one database are not consistent for all other different databases. A method showing high performance in one database might not show same high performance in other database. Since, we conducted our experiments on the retina images from various sources and all images are of different size and quality; we can say that the proposed method for CRVO detection is a versatile method whose performance should be consistent for any database of retinal fundus image. Therefore, it is feasible to implement in real.

5.1 The Basic of the Convolutional Neural Network (CNN)

The Convolutional Neural Network is the advanced version of the general Neural Network (NN); used in various areas, including image and pattern recognition, speech recognition, natural language processing, and video analysis [12]. The CNN facilitates the deep learning to extract abstract features from the raw image pixels.

CNNs take a biological inspiration from the visual cortex. The visual cortex has lots of small cells that are sensitive to specific regions of the visual field, called the receptive field. This small group of cells functions as local filters over the input space. This idea was expanded upon by Hubel and Wiesel, where they showed that some individual neuronal cells in the brain responded (or fired) only in the presence of edges of a certain orientation. For example, some neurons fired when exposed to vertical edges and some when shown horizontal or diagonal edges. They found out that all of these neurons were structured as a columnar architecture and are able to produce visual perception [13]. This idea of specialized components inside of a system having specific tasks (the neuronal cells in the visual cortex looking for specific characteristics) is one that machines use as well, and is the basis behind

CNNs. By assembling several different layers in a CNN, complex architectures are constructed for classification problems. The CNN architecture consists of four types of layers: convolution layers, pooling/subsampling layers, non-linear layers, and fully connected layers [13, 15].

5.1.1 The Convolutional Layer

The first layer in a CNN is always a Convolutional Layer. The convolution functions as feature extractor that extracts different features of the input. The first convolution layer extracts the low-level features like edges, lines, and corners. Higher-level layers extract the higher-level features. Suppose, the input is of size $M \times M \times D$ and is convolved with K kernels/filters, each of size $n \times n \times D$ separately. Convolution of an input with one kernel produces one output feature. Therefore, the individual convolution with K kernels produces K features. Starting from top-left corner of the input, each kernel is moved from left to right and top to bottom until the kernel reaches the bottom-right corner. For each stride, element-by element multiplication is done between $n \times n \times D$ elements of the input and $n \times n \times D$ elements of the kernel on each position of the kernel. So, $n \times n \times D$ multiply-accumulate operations are required to create one element of one output feature [13, 15].

5.1.2 The Pooling Layer

The pooling layer reduces the spatial size of the features. It makes the features robust against noise and distortion. It also reduces the number of parameters and computation. There are two ways for down sampling: max pooling and average pooling. Both the pooling functions divide the input into non-overlapping two dimensional space [12].

5.1.3 Non-Linear layer

The non-linear layer adds non linearity to the network as the real world data are non-linear in nature (https://ujjwalkarn.me/2016/08/11/intuitive-explanation-convnets/). The rectified linear unit (ReLU) is a nonlinear layer that triggers a certain function to signal distinct identification of likely features on each hidden layer. A ReLU performs the function y = max(x,0) keeping the output size same as the input. It also helps to train faster.

5.1.4 Fully Connected Layers

In a CNN, the last final layer is a fully connected layer. It is a Multi-Layer Perceptron (MLP) that uses an activation function to calculate the weighted sum of all the features of the previous layer to classify the data into target classes.

5.2 Methodology

Some of the popular Convolutional Networks are LeNet, AlexNet, ZF Net, GoogLeNet, VGGNet and ResNet. The LeNet is the first successful Convolutional Network used for recognizing digits, zip codes etc. After LeNet, AlexNet came as a deeper version of LeNet which was successfully used for object recognition in large scale. ZF Net is modified version of the AlexNet where the hyper parameters are modified. The GoogLeNet introduced inception module to drastically reduce the number of the parameters. VGGNet is a large deep Convolutional Network with 16 Convolutional and Fully Connected layers. ResNet skipped the fully connected layers and made heavy use of batch normalization. Moreover, some CNNs are fine tuned or the architecture is tweaked for different applications. For e.g. in [16], the authors designed a CNN for facial landmark detection. Again in [17] and [18], the basic CNN is fined tuned to identify different EEG signals. Our designed Convolutional Network is based on LeNet architecture. The general structure of LeNet is as follows:

Input=>Conv=>Pool=>Conv=>Pool=>FC=>ReLu=>FC=>Output

Our designed CNN structure is as follows:

Input=>Conv=>ReLU=>Pool=>Conv=>ReLU=>Pool=>Conv=>Re

LU=>Pool=>FC=>ReLU=>FC=>Output

Before feeding the retina image to the CNN, we performed preprocessing to enhance the quality of the image. Since, we have collected the color fundus image of normal retina and the retina with CRVO images from multiple databases, all the images are of different sizes and of different formats. The images from the STARE databases are of size 700×605 and TIF format. The images from the DRIVE database are 565×585 TIFF images. The images from the Dr. Hossein Rabbani are 1612×1536 JPEG images. The each of the images from the Retina Image Bank is of different sizes and format. We converted all the images to TIF format and resized into a standard image size 60×60.

5.2.1 Image Preprocessing

After converting all the images into 60×60 TIF format, we extracted the green channel as it provides the distinct visual features of the retina compared to other two (Red and Blue) channels. Then, an Average filter of size 5×5 is applied to remove the noise. After that the contrast of that grayscale retina image is enhanced by applying Contrast-limited adaptive histogram equalization (CLAHE). CLAHE operates on small regions in the image and enhance the contrast of each small region individually. A bilinear interpolation is used to combine the neighboring small regions in order to eliminate artificial boundaries. The contrast of the homogeneous areas can be limited to avoid unwanted noise present in the image. Figure 5a shows the normal image and Fig. 5b shows the green channel of the RGB image, Fig. 5c

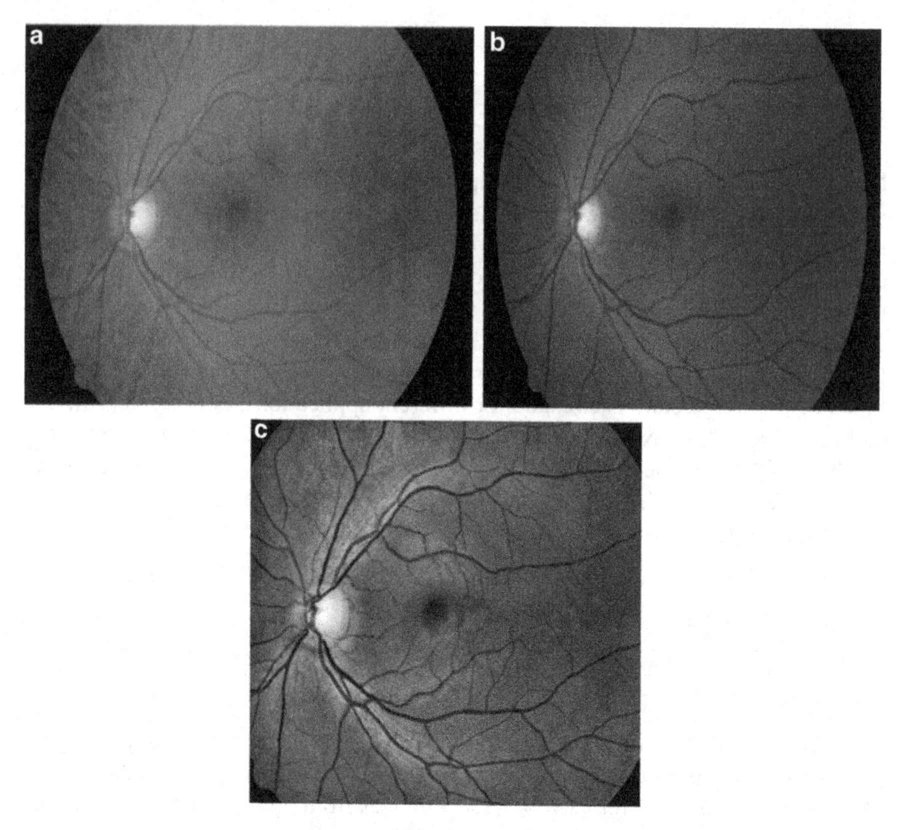

Fig. 5 (**a**) Normal *RGB images*, (**b**) *Green* channel. (**c**) Pre-processed image

shows the enhanced image. Figure 6a shows the CRVO image, Fig. 6b shows the green channel and Fig. 6c shows the enhanced image.

5.2.2 The Network Topology

The designed CNN for recognition of CRVO consists of 12 layers, including three convolution layers, three pooling layers, four ReLUs and two fully connected layers. We have two classes: Normal image and CRVO image. The layers in the CNN network are stacked with three sets of convolution layer, followed by ReLU followed by a pooling, followed by a fully connected layer, ReLU and fully connected layer. Finally, the features obtained by the 4th ReLU layer are transferred to the last fully connected layer. Ultimate classification of CRVO is based on these high level features. Softmax function is used as the activation function. The network topology is shown in Fig. 7.

Layer 1: The first convolutional layer convolves the input retina image of size 60×60. In the first layer, we used 32 filters of size 5×5 with a stride of 1 to outputs

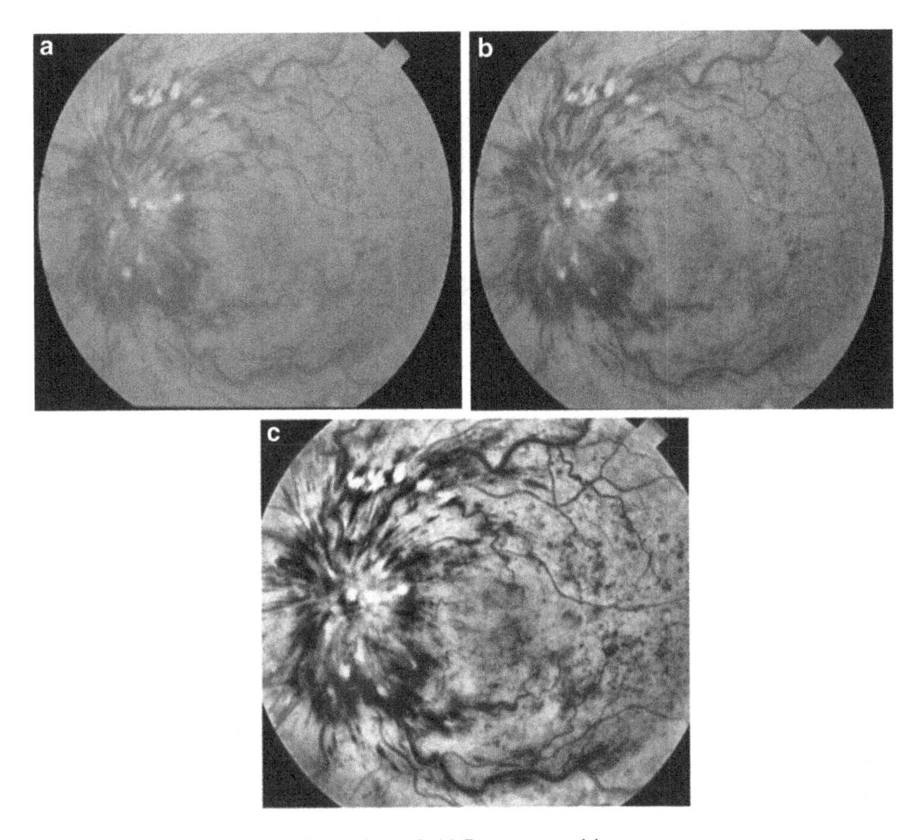

Fig. 6 (**a**) CRVO image, (**b**) *Green* channel, (**c**) Pre-processed image

the feature data map. Mathematically, the operation of a convolution layer can be formulated as follows:

$$y_j^n = f\left(\sum_{i\epsilon N_j} W_{ij}^n y_i^{n-1} - b_j^n\right) \quad (3)$$

where y_i^{n-1} is the input of convolution layer. W_{ij}^n is a convolution kernel weight of layer n with the size of $i \times j$. b_j^n is a bias, y_j^n is the output and N is the number of inputs used to generate y_j^n. After the first convolution layer the output feature map is of size $56 \times 56 \times 32$.

Layer 2: After the first convolution layer, a rectified non-linear unit (ReLU) is used. It increases the nonlinear property keeping the output volume same as the input volume.

Layer 3: The output feature map of ReLU is given as input to a pooling layer. For the pooling layer, we used max pooling to reduce the size of the output feature map and capture the spatial information. A filter size 2×2 and stride 2 are used for the max pooling. The equation of pooling layer can be given by,

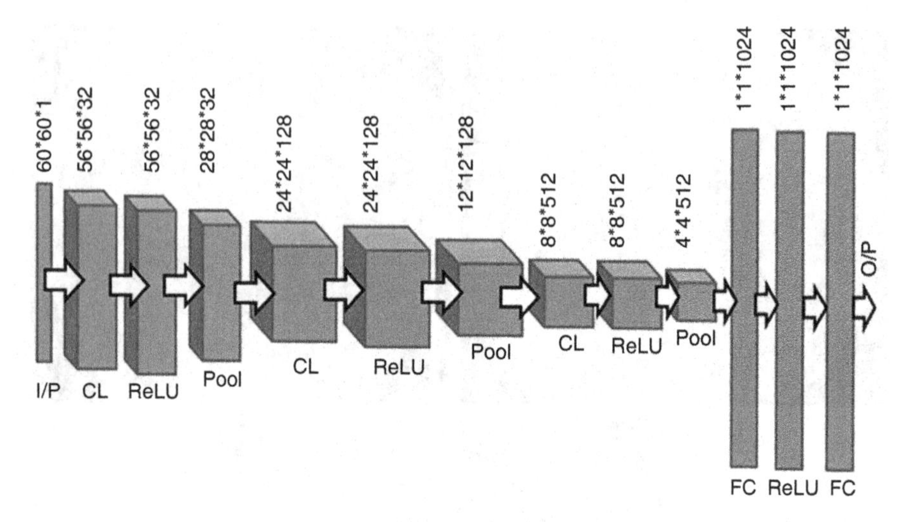

Fig. 7 The CNN topology

$$x_j^l = f\left(\beta_j^l down\left(x_i^{l-1}\right) + b_j^l\right) \tag{4}$$

where function *down(.)* denotes a max pooling function for our network. β_j^l is a weight and b_j^l is bias. The equation for max pooling function with a filter dimension $m \times m$ can be given by,

$$y = \max\left(x_i\right), i \in \{1, 2, \ldots, m \times m\} \tag{5}$$

After max-pooling we get an output feature volume $28 \times 28 \times 32$.

Layer 4: The output of the pooling layer is fed to the 2nd convolution layer. With 128 filters of size 5×5 we get an output activation map $24 \times 24 \times 128$.

Layer 5: With the 2nd ReLU the nonlinear properties are further increased keeping the output volume same.

Layer 6: The 2nd max-pooling further reduces the number of features to an output volume$12 \times 12 \times 128$.

Layer 7: In the 3rd convolution layer the output of pooling layer is convolved with 512 filters of dimension 5×5 to get output activation $8 \times 8 \times 512$.

Layer 8: The ReLU changes the negative activation to 0 to further increase the nonlinearity.

Layer 9: The max-pooling down samples the input of ReLU to output volume 4×4 with receptive field size 2×2.

Layer 10: This layer is a fully connected layer converted to a convolutional layer with a filter size 4×4 with 1024 kernels. It generates an activation map of size $1 \times 1 \times 1024$

Layer 11: The ReLU enhances the non-linear property.

Layer 12: The output of the ReLU is fed to another fully connected layer. From a single vector 1024 class scores, 2 classes: Normal and CRVO images are classified.

6 Result and Discussion

For our experiment, we collected 108 CRVO images (26 images from STARE database and 84 images from Retina Image Bank) and 133 Normal images (100 images from Hossein Rabbani database, 30 images from STARE database and 3 images from DRIVE database). We trained the CNN network with 100 normal (randomly selected from normal images from Hossein Rabbani, STARE and DRIVE databases) and 100 CRVO grayscale images (randomly selected from STARE and Retina Image Bank's CRVO images) of size 60×60 after preprocessing. In the 1st, 2nd and 3rd convolution layer, the filter size is 5×5 and in the 4th or last convolution layer/fully connected layer the filter size is 4×4. The numbers of filters or kernels in the four convolution layers are 32, 128, 512 and 1024 respectively. The training is done with an epoch 70. For each training epoch we provided a batch size of nine training images and one validation image. Using the designed classifier we obtained an accuracy of 97.56%. Figure 8 shows the network training and validation for epoch 70. Figure 9 shows the Cumulative Match Curve (CMC) for rank vs. recognition rate. For the two classes we tested 41 images, 8 test images

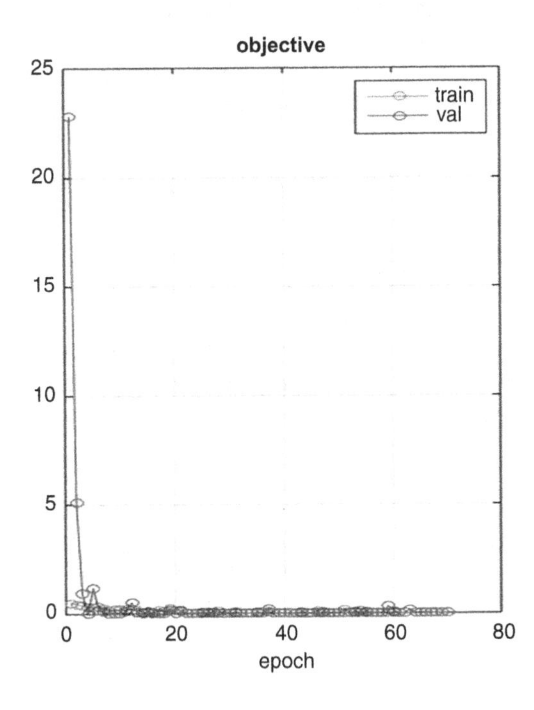

Fig. 8 Network Training for epoch 70

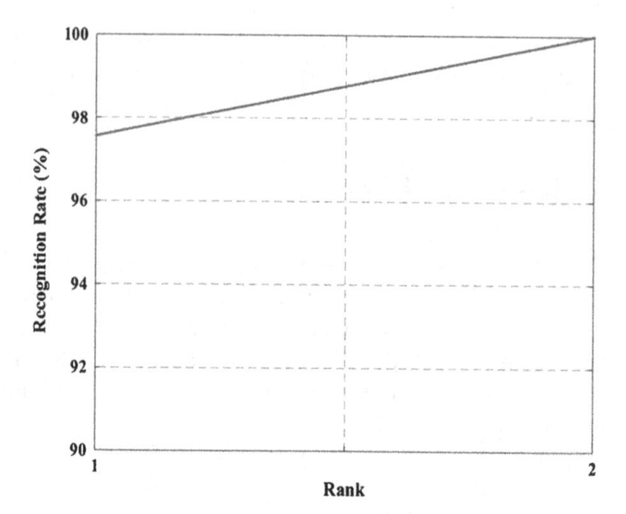

Fig. 9 Cumulative Match Curve (CMC)

for CRVO and 33 test images for normal retina. Each test produces a score for each image while comparing to each target class. If the score between test image and one of the target classes is larger than the other class, then that class is recognized in the first rank. Here, out of 41 test images 40 images are correctly recognized, hence the recognition rate for rank 1 is 97.56% and for rank 2 recognition rate is 100%.We further evaluated the performance in terms of specificity, sensitivity, positive predictive value and negative predictive value. Sensitivity is the probability of the positive test given that the patient has the disease. It measures the percentage of the people actually having the disease diagnosed correctly. Sensitivity can be given by following equation:

$$Sensitivity = \frac{True\ Positive}{True\ Positive + False\ Negative} \tag{6}$$

where, the "*True Positive*" depicts correctly identified disease and "*False Negative*" describes incorrectly rejected people having disease. In our experiment we got sensitivity 1. That means all the CRVO images are detected correctly. Again, specificity is the probability of a negative test given that the patient has no disease. It measures the percentage of the people not having disease diagnosed correctly. Specificity can be given by following equation:

$$Specificity = \frac{True\ Negative}{True\ Negative + False\ Positive} \tag{7}$$

In our experiment, one normal image is incorrectly detected as CRVO image; hence, we obtained the specificity of 0.9697. The Positive Predictive value is the probability that subjects with a positive screening test truly have the disease. We got a positive predictive value 0.889. The Negative Predictive value is the probability

Table 2 Performance evaluation of the system

Accuracy	Sensitivity	Specificity	Error rate	Positive predictive value	Negative predictive value
97.56%	1	0.9697	2.44	0.8889	1

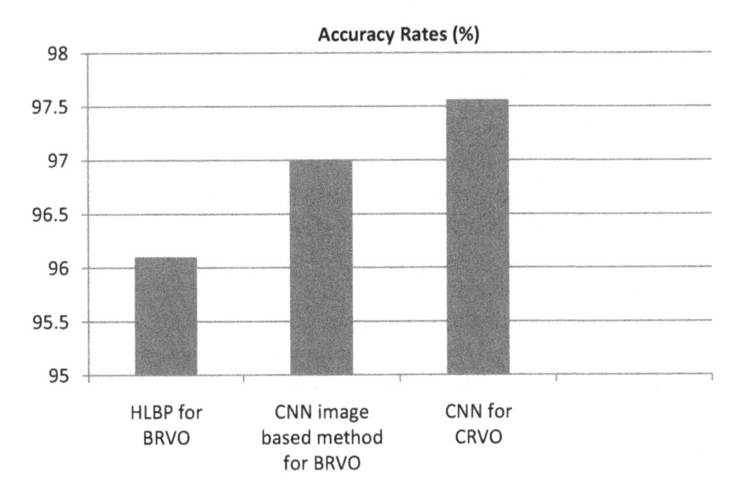

Fig. 10 Accuracy rates of different methods for BRVO and proposed method for CRVO

that subjects with a negative screening test truly do not have the disease. We obtained negative predictive value 1. Table 2 summarizes the total evaluation of the system.

The experimental results show that the proposed method of detecting CRVO using CNN is a powerful method that we can implement in practice. Since there is no existing automatic detection of CRVO found in the literature, we are the first group to work on the automatic recognition of CRVO. Therefore, it is also difficult to compare the results with other methods. However, if we compare the method with that of automated recognition of BRVO, then our method performs better than the other feature extraction techniques and slightly better than the CNN based method. Figure 10 shows the comparison of our method with the existing methods for automated recognition of BRVO. So, the proposed method is fulfilling the need of automatic detection of CRVO to help the ophthalmologists in faster and efficient diagnosis of CRVO. It will also save the time and money of the patients. The method is taking care of the problems related to the image quality. This method is handling the most important issue, i.e., the requirement of different segmentation and feature extraction methods for detecting the abnormalities appear due to CRVO. Especially in the case, when detecting flame shaped haemorrhages, dilated veins and tortuous veins in the early stage of CRVO could be complicated and computationally expensive task. The supreme performance of the proposed CNN based method with a correct accuracy rate of 97.57% for the images of different sources proves it to be a promising consistent system for the automatic detection of CRVO. Because, all the images are captured by different funduscope devices and have different types (image captured in different angles), format, resolution and quality. The performances of the existing automatic detection systems for BRVO are limited to a single dataset consisting of same type (image captured in same angle), same quality and same resolution images. Therefore, this CNN method is an efficient, versatile and consistent method for detecting CRVO.

7 Conclusion

In this chapter, we proposed a Central Retinal Vein Occlusion (CRVO) recognition method using Convolutional Neural Network (CNN). The designed network takes grayscale preprocessed images and recognizes the retina image with CRVO and the normal retina image. We have achieved a high accuracy of 97.56%. The proposed method is an image based method which is quite practical to implement. The advantage of this system is that there is no requirement of extra feature extraction step. The convolution layer serves both as feature extractor and the classifier. It is difficult to design feature extraction algorithm for the clinical signs of CRVO. Because, most of the time CRVO affects the whole retina and those large size hemorrhages, cotton wool spots are hard to define by other feature extraction methods. In CNN, each convolution layer extracts the low level features to the high level features from the CRVO images. Hence, it saves time. Since we conducted the experiment on retina image from different sources, the general automated detection method might affect the accuracy of the overall system due to different image quality, size and angle. However, use of CNN handles this situation due to its ability to cope with the distortions such as change in shape due to camera lens, different lighting conditions, different poses, presence of partial occlusions, horizontal and vertical shifts, etc. Therefore, the proposed CRVO detection scheme is a robust method.

References

1. The Royal College of Ophthalmologists: Retinal Vein Occlusion (RVO) Guidelines. The Royal College of Ophthalmologists, London (2015)
2. Sperduto, R.D., Hiller, R., Chew, E., Seigel, D., Blair, N., Burton, T.C., Farber, M.D., Gragoudas, E.S., Haller, J., Seddon, J.M., Yannuzzi, L.A.: Risk factors for hemiretinal vein occlusion: comparison with risk factors for central and branch retinal vein occlusion: the eye disease case-control study. Ophthalmology. **105**(5), 765–771 (1998)
3. Hayreh, S.S.: Prevalent misconceptions about acute retinal vascular occlusive disorders. Prog. Retin. Eye Res. **24**, 493–519 (2005)
4. Central Retinal Vein Occlusion Study Group: Natural history and clinical management of central retinal vein occlusion. Arch. Ophthalmol. **115**, 486–491 (1997)
5. Jelinek, H., Cree, M.J.: Automated image detection of retinal pathology. CRC Press, Boca Raton (2009)

6. Noma, H.: Clinical diagnosis in central retinal vein occlusion. J. Med. Diagn. Methods. **2**, 119 (2013)
7. Shiraishi, J., Li, Q., Appelbaum, D., Doi, K.: Computer-aided diagnosis and artificial intelligence in clinical imaging. Semin. Nucl. Med. **41**(6), 449–462 (2011)
8. Zhang, H., Chen, Z., Chi, Z., Fu, H.: Hierarchical local binary pattern for branch retinal vein occlusion recognition with fluorescein angiography images. IEEE Electr. Lett. **50**(25), 1902–1904 (2014)
9. Gayathri, R., Vijayan, R., Prakash, J.S., Chandran, N.S.: CLBP for retinal vascular occlusion detection. Int. J. Comput. Sci. Iss. **11**(2), 204–209 (2014)
10. Fazekas, Z., Hajdu, A., Lazar, I., Kovacs, G., Csakany, B., Calugaru, D.M., Shah, R., Adam, E.I., Talu, S.: Influence of using different segmentation methods on the fractal properties of the identified retinal vascular networks in healthy retinas and in retinas with vein occlusion. In: 10th Conference of the Hungarian Association for Image Processing and Pattern Recognition (KÉPAF 2015), pp. 360–373 (2015)
11. Zhao, R., Chen, Z., Chi, Z.: Convolutional neural networks for branch retinal vein occlusion recognition. In: IEEE International Conference on Information and Automation (2015)
12. Hijazi, S., Kumar, R., Rowen, C.: IP group cadence. https://ip.cadence.com/uploads/901/cnn_wp
13. LeNet details description: http://deeplearning.net/tutorial/lenet.html
14. Orth, D.H., Patz, A.: Retinal branch vein occlusion. Surv. Ophthalmol. **22**, 357–376 (1978)
15. A Beginner's Guide to Understanding Convolutional Neural Networks-Adit Deshpande of UCLA: https://adeshpande3.github.io/adeshpande3.github.io/A-Beginner's-Guide-To-Understanding-Convolutional-Neural-Networks/
16. Wu, Y., Hassner, T., Kim, K., Medioni, G., Natarajan, P.: Facial landmark detection with tweaked convolutional neural networks. Comput. Vision Pattern Recogn. **2**, preprint arXiv:1511.04031 (2015)
17. Ma, L., Minett, J.W., Blu, T, Wang, W.S: Resting state EEG-based biometrics for individual identification using convolutional neural networks. In: 37th Annual International Conference of the IEEE Engineering in Medicine and Biology Society (EMBC) (2015)
18. Cecotti, H., Graser, A.: Convolutional neural networks for P300 detection with application to brain-computer interfaces. IEEE Trans. Pattern Anal. Mach. Intell. **33**(3), 433–445 (2011)

Swarm Intelligence Applied to Big Data Analytics for Rescue Operations with RASEN Sensor Networks

U. John Tanik, Yuehua Wang, and Serkan Güldal

Abstract Various search methods combined with frontier technology have been utilized to save lives in rescue situations throughout history. Today, new networked technology, cyber-physical system platforms, and algorithms exist which can coordinate rescue operations utilizing swarm intelligence with Rapid Alert Sensor for Enhanced Night Vision (RASEN). We will also introduce biologically inspired algorithms combined with proposed fusion night vision technology that can rapidly converge on a near optimal path between survivors and identify signs of life trapped in rubble. Wireless networking and automated suggested path data analysis is provided to rescue teams utilizing drones as first responders based on the results of swarm intelligence algorithms coordinating drone formations and triage after regional disasters requiring Big Data analytic visualization in real-time. This automated multiple-drone scout approach with dynamic programming ability enables appropriate relief supplies to be deployed intelligently by networked convoys to survivors continuously throughout the night, within critical constraints calculated in advance, such as projected time, cost, and reliability per mission. Rescue operations can scale according to complexity of Big Data characterization based on data volume, velocity, variety, variability, veracity, visualization, and value.

Keywords Autonomous • Drones • Swarm intelligence • Night vision • RASEN • Rescue operations • Big Data • Visualization

U.J. Tanik (✉) • Y. Wang
Department of Computer Science, Texas A&M University-Commerce, Commerce, TX, USA
e-mail: john.tanik@tamuc.edu; yuehua.wang@tamuc.edu

S. Güldal
Department of Computer Science, University of Alabama at Birmingham, Birmingham, AL, USA
e-mail: guldal@uab.edu

S.C. Suh, T. Anthony (eds.), *Big Data and Visual Analytics*,
https://doi.org/10.1007/978-3-319-63917-8_2

1 Introduction

Historically, optimization methods combined with frontier technology have been utilized to save lives in rescue situations. Today, new technology and search algorithms exist which can optimize rescue operations utilizing Rapid Alert Sensor for Enhanced Night vision (RASEN). We will introduce biologically inspired algorithms combined with fusion night vision technology that can rapidly converge on optimal paths for discovering disaster survivors and the rapid identification of signs of life for survivors trapped in rubble. Networked data visualization is provided to rescue teams based upon swarm intelligence sensing results so that appropriate relief supplies can optimally be deployed by convoys to survivors within critical time and resource constraints (e.g. people, cost, effort, power).

Many countries have rescue strategies in development for disasters like fires, earthquakes, tornadoes, flooding, hurricane, and other catastrophes. In 2016, the world suffered the highest natural disaster losses in 4 years, and losses caused by disasters worldwide hit $175 billion [1]. Search and rescue (SAR) missions are the first responder for searching for and providing relief to people who are in serious and/or imminent danger. Search and rescue teams and related support organizations take actions for searching and rescuing victims from varying incident environments and locations. During search and rescue, lack of visibility, especially at night, has been considered one of the major factors affecting rescue time and therefore, rescue mission success. Poor night visibility and diverse weather conditions also makes searching, detecting, and rescuing more difficult and sometimes even impossible if survivors are hidden behind obstacles. Furthermore, poor visibility is also a common cause of roadway accidents given that vision provides over 90% of the information input used to drive [2]. In fact, it has been reported that the risk of an accident at night is almost four times greater than during the day [3]. When driving at night our eyes are capable of seeing in limited light with the combination of headlights and road lights, however, our vision is weaker and more blurry at night, adding difficulty when avoiding moving objects that suddenly appear.

Recent years have seen significant advancement in the fields of mobile, sensing, communications, and embedded technologies, and reduction in cost of hardware and electronic equipment. This has afforded new opportunities for extending the range of intelligent night vision capabilities and increasing capabilities for searching and detecting pedestrians, vehicles, obstacles, and victims at night and under low light conditions.

Herein, intelligent physical systems are defined to be machines and systems for night vision that are capable of performing a series of intelligent operations based upon sensory information from cameras, LIDAR, radar and infrared sensors in complex and diverse Big Data analytic environments. These intelligent machines can be used for various applications, including power line inspection, automotive, construction, precision agriculture, and search and rescue, which is the focus of this chapter. Each application requires varying levels of visibility. Unlike traditional systems which only have a single purpose or limited capabilities and require

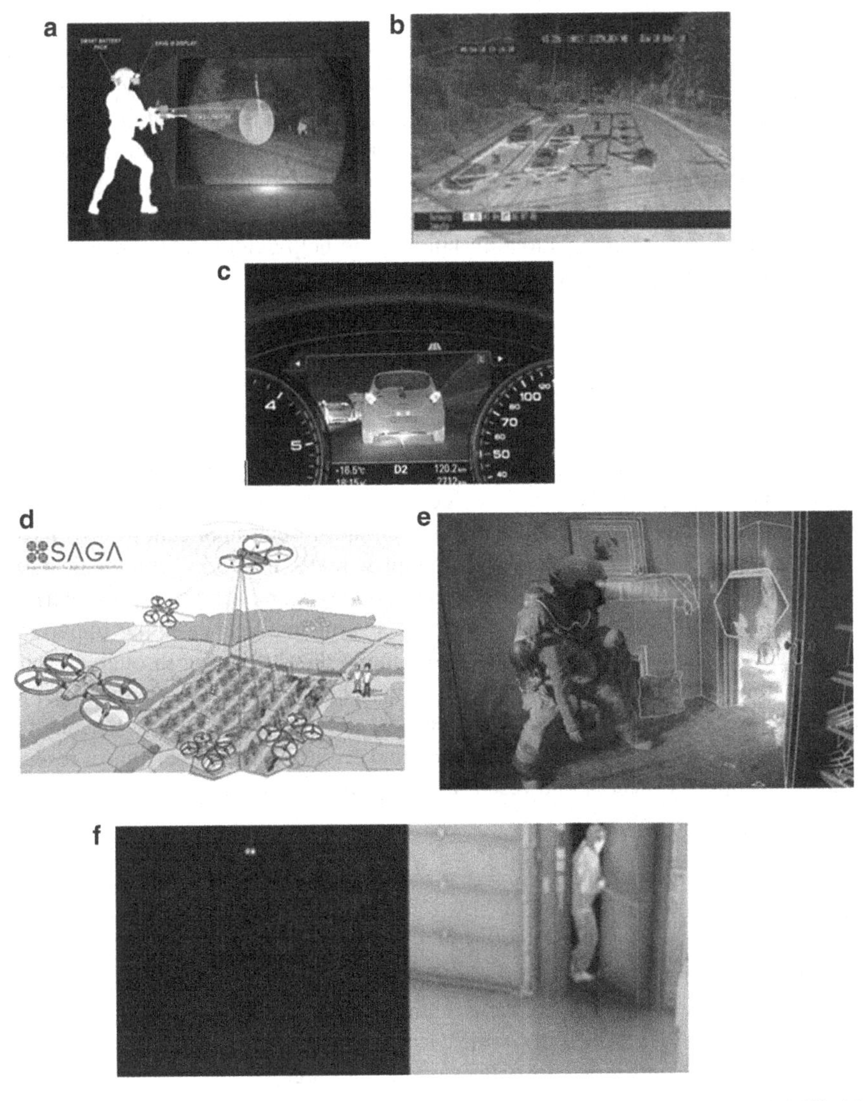

Fig. 1 Sample intelligent physical systems. (**a**) ENVG II [4]. (**b**) Traffic control (FLIR) [5]. (**c**) Automotive [6]. (**d**) Precision agriculture (SAGA) [7]. (**e**) Firefighting (C-Thru) [8]. (**f**) Security (ULIS) [9]

human intervention during missions, intelligent physical systems which include night vision, combines computing, sensing, communication, and actuation, in order to tailor operational behavior in accordance with a particular collected operational information context. Figure 1 depicts six sample intelligent physical systems for potential use during night or low-light vision conditions.

Advantages of night vision based intelligent physical systems are their ability to sense, adapt and act upon changes in their environments. Becoming more aware of the detailed operational context is one important requirement of night vision based intelligent physical systems. As every domain application is different, it is difficult to provide a single system or technique which provides a solution for all specialized needs and applications. Therefore, our motivation is to provide an overview of night vision based intelligent machine systems, and related challenges to key technologies (e.g. Big Data, Swarm, and Autonomy) in order to help guide readers interested in intelligent physical systems for search and rescue.

2 Literature Survey

Efficient communication and processing methods are of paramount importance in the context of search-and-rescue due to massive volume of collected data. As a consequence, in order to enable search-and-rescue applications, we have to implement efficient technologies including wireless networks, communication methodologies, and data processing methods. Among them, Big Data (also referred to as "big data"), artificial intelligence, and swarm intelligence allow important advantages to real-time sensing and large-volume data gathering through search-and-rescue sensors and environment. Before elaborating further on the specific technologies fitting into the search and rescue scenarios, we outline the unique features of rescue drones, review challenges, and discuss potential benefits of rescue drones in supporting search-and-rescue applications.

2.1 Rescue Drones

Drones, commonly known as Unmanned Aerial Vehicles (UAV), are small aircraft which perform automatically without human pilots. They could act as human eyes and can easily reach areas which are too difficult to reach or dangerous for human beings and they can collect images through aerial photography [12]. Compared to skillful human rescuers (e.g., police helicopter, CareFlite etc.) and ground based rescue robots, the use of UAVs in emergency response and rescue has been emerging as a cost-effective and portable complement for conducting remote sensing, surveying accident scenes, and enabling fast rescue response and operations, as depicted in Fig. 2. A drone is typically equipped with a photographic measurement system, including, but not limited to, video cameras, thermal or infrared cameras, airborne LiDAR (Light Detection and Ranging) [13], GPS, and other sensors (Fig. 3). The thermal or infrared cameras can be particularly useful for detecting biological organisms such as animals and human victims and for inspecting inaccessible buildings, areas (e.g. Fukushima), and electric power lines.

Fig. 2 Rescue scenario with drones [10]

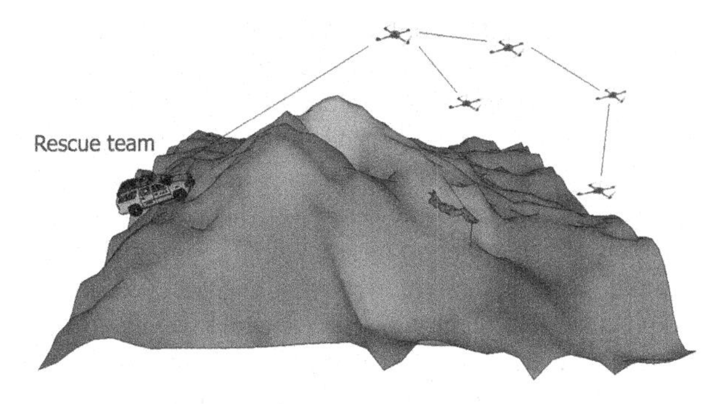

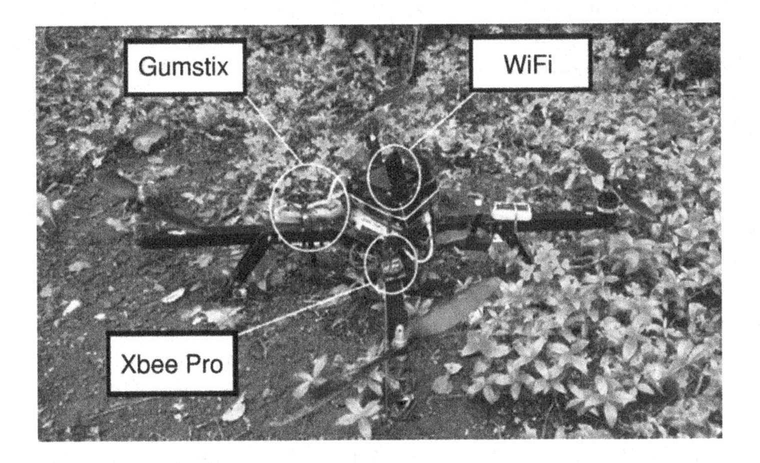

Fig. 3 Flying unit: Arducopter [11]

Airborne LiDAR can operate day and night and is generally used to create fast and accurate environmental information and models.

Drones are ideal for searching over vast areas that required Big Data analytics; however, drones are often limited by factors such as flying time and payload capacity. Many popular drones on the market need to follow preprogrammed routes over a region and can only stay airborne for a limited period of time. This limitation has increased research conducted for drone-aided rescue. The research includes path planning [14, 15], aerial image fusion [12, 16–19], and drone swarm [20, 21].

Early research has focused on route path planning problems in SAR motivated by minimizing time from initial search to rescue which can range from hours, days, to even months after the disaster. Search efficiency affects the overall outcome of SAR, so that the time immediately following the event requires a fast response in order to locate survivors on time. The path planning is generally used to find a collision-free flight path and to cover maximum area in adverse environments in the presence of static and dynamic obstacles under various weather conditions with minimal user intervention. The problem is not simply an extension or variation of UAV path planning aiming to find a feasible path between two points [22, 23]. For example, the complete-coverage method, local hill climbing scheme, and evolutionary algorithms, developed by Lin and Goodrich [14] defined the problem as a discretized combinatorial optimization problem with respect to accumulated probability in the airspace. To reduce the complexity of the path planning problem, the study [15] divided the terrain of the search area into small search areas, each of which was assigned to an individual drone. Each drone initializes its static path planning using a Dijkstra algorithm and uses Virtual Potential Function algorithm for dynamic path planning with a decentralized control mechanism.

Aerial images, infrared images, and sensing data captured by drones enable rescue officers and teams to have a more detailed situational awareness and increased comprehensive damage assessment. Dong et al. [17] presented a fast stereo aerial image construction method with a synchronized camera-GPS imaging system. The high precision GPS is used to pre-align and stitch serial images. The stereo images are then synthesized with pair-wise stitched images. Morse et al. [18] created coverage quality maps by combining drone-captured video and telemetry with terrain models. The facial recognition is another task of great interest. Hsu and Chen [12] compared the use of aerial images in face recognition so as to identify specific individuals within a crowd. The focus of the study [19] lies on real-time vision attitude and altitude estimation in low light or dark environments by means of a combination of camera and laser projector.

Swarm behavior of drones is featured by coordinated functions of multiple drones, such as collective decision making, adaptive formation flying, and self-healing. Drones need to communicate with each other to achieve coordination. Burkle et al. [20] refined the infrastructure of drone systems by introducing a ground central control station as a data integration hub. Drones can not only communicate with each other, but also exchange information with the ground station to increase optimization of autonomous navigation. Gharibi et al. [21] investigated layered network control architectures for providing coordination for efficiently utilizing the controlled airspace and providing collision-free navigation for drones. Rescue drones also need to consider networking described next.

2.2 Drone Networking

In typical scenarios, drones fly over an area, perform sensory operations, and transmit gathered information back to a ground control station or the operation center via networks (Figs. 4 and 5). However, public Internet communication networks are often unavailable or broken in remote or disaster areas. The question that arises now is how to find a rapid, feasible way of re-establishing communications, while remaining connected to the outside world for disaster areas. The rise of rescue drones and extensive advancements in communication and sensing technologies drives new opportunities in designing feasible solutions for the communication problem. Besides data gathering, rescue drones can act as a temporary network

Fig. 4 MQ-9 reaper taxiing [24]

Fig. 5 Airnamics R5

access points for survivors and work cooperatively to forward and request data back to the ground control station [10, 11, 25–27].

In the literature, there are two types of rescue drone network systems: single-drone and multiple-drone. The single drone network system generally has a star topology, in which drones are working independently and linked to a ground control station. In [11], drones are equipped with WiFi (802.11n) module and responsible for listening to survivor "HELP" requests in communication range. The drone then forwards the "HELP" request to the ground control station through an air-to-ground communication link that is a reliable, IEEE 802.15.4-based remote control link with low bandwidth (up to 250 kbps) but long communication range (up to 6 km), as included in Table 1 [28], which also used a single drone and developed a contour map based location strategy for locating targets. However, the outcome and efficiency of search and rescue are greatly restricted by single drone systems, where the single drone [24] can only have limited amount of coverage increases.

Instead of having only one (large or heavy-lift) drone in the system, multiple drones are deployed, working interactively for sensing and transmitting data in

Table 1 Existing wireless technologies for drone communication

	XBee (product name)	ZigBee	Bluetooth	WiFi	WiMAX	LTE-advanced
Standard	IEEE802.15.4 and ZigBee compliance	IEEE 802.15.4	IEEE 802.15.1	IEEE 802.11	IEEE 802.16	UMTS/4GSM
Data rate	Up to 250 kbps	Up to 250 kbps	Up to 1 Mbps	11 and 54 Mbps	5–75 Mbps	Up to 3 Gbps
Range	5 km	10–150 m	Up to 10 m	Up to 100 m	Up to 50 km	Up to 100 km
Networking topology	P2P, star, or mesh	P2P, star, or mesh	Star	Star	Star	Star
Frequency	2.400–2.483 GHz	2.4 GHz, 868 and 915 MHz	2.4 GHz	2.4, 3, 5 and 60 GHz	2.3–2.5 and 3.4–3.5 GHz	Licensed band
Typical applications	Remote monitoring and control	Remote monitoring and control	Short-range communication and inter-device wireless connectivity	WLAN and Ad-hoc network	Backhaul, mobile internet, last mile access, and rural connectivity	VoIP, video/audio streaming, real-time gamming
Power consumption	Very low	Very low	Low	High	High	High

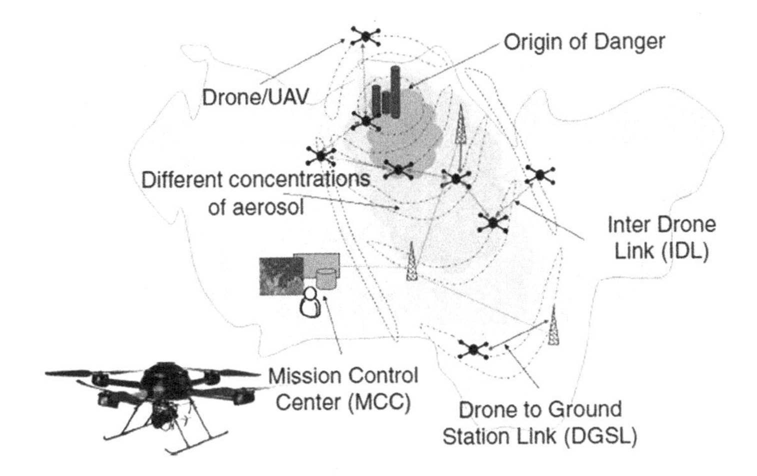

Fig. 6 Air shield [25]

multiple-drone systems [25, 26, 29–31], as shown in Figs. 6 and 7. Generally, the system is composed of multiple drones and a ground control center. The drones are small or middle-sized unmanned aerial vehicles equipped with wireless transceivers, GPS, power supply systems, and/or on-board computers. The wireless transceivers are modules to provide wireless end-point connectivity to drones. The module can use xBee, ZigBee, WiFi, Blue-tooth, WiMAX, and LTE protocols for fast or long distance networking. Table 1 shows available wireless communication technologies for drone systems. In particular, each technology has its unique characteristics and limitations to fulfill the requirements of drone networks. Bluetooth and WiFi technology are main short-range communication technologies and generally used to build small wireless ad-hoc networks of drones. The communication links allow drones to exchange status information with each other during networked flight.

Daniel et al. [25] used this idea and built a multi-hop drone-to-drone (mesh) and single-hop drone-to-ground network. Given not all drones have a connection to the ground control station, the inter-drone links guide data routing towards the station. This process repeats until the data reaches a drone with drone-to-ground link realized with wireless communication techniques WiMAX and LTE. Cimino et al. [32] claimed that WiMAX can also be used for inter-drone communication. SAR Drones [26] studied the squadron and independent exploration schemes of drones. Drones can also be linked to satellites in multi-drone systems [21, 33].

It is possible that drones might fly outside of the communication range of the ground communication system, as shown in Fig. 7. PhantomPilots: Airnamics [29] proposed a multi-drone real-time control scheme based on multi-hop Ad-hoc networking. Each drone acts as a node of the Ad-hoc network and uses the ad- hoc network to transmit the data to the ground control station via drones in the station communication range.

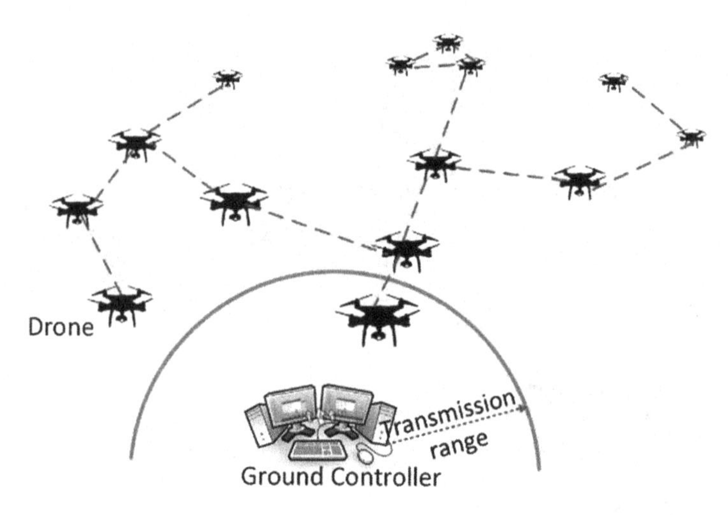

Fig. 7 Multi-drone control system [29]

Beside single-layer networks, there are also dedicated multi-layer networks designed for multi-drone systems. Asadpour et al. [34] proposed a 2-layer multi-drone network, as shown in Fig. 8. Layer I consists of airplanes (e.g. Swinglet in Fig. 8a) which are employed to form a stable, high-throughput wireless network for copters (e.g. Arducopter in Fig. 8b). Copters are at layer II to provide single-hop air-to-ground connection for victims and rescue teams. For efficient search and rescue, controlled mobility can be applied to airplanes and copters to maximize network coverage and link bandwidth. In [35], three categories of drones: blimps, fixed wing, and vertical axis drones were considered to constitute a multi-layer organization of the drone fleet with instantaneous communication links. Big Data in rescue operations introduces another factor of complexity.

2.3 Regional Disasters

Night vision systems for search and rescue are undergoing a revolution driven by the rise of drones and night vision sensors to gather data in complex and diverse environments and by the use of data analytics to guide decision-making. Big Data collecting from satellites, drones, automotive, sensors, cameras, and weather monitoring all contain useful information about realistic environments. The complexity of data includes consideration of data volume, velocity, variety, variability, veracity, visualization, and value. The ability to process and analyze this data to extract insight and knowledge that enable in-time rescue, intelligent services, and new ways to assess disaster damage, is a critical capability. Big Data analytics is actually not a new concept or paradigm. However, in addition to cloud computing,

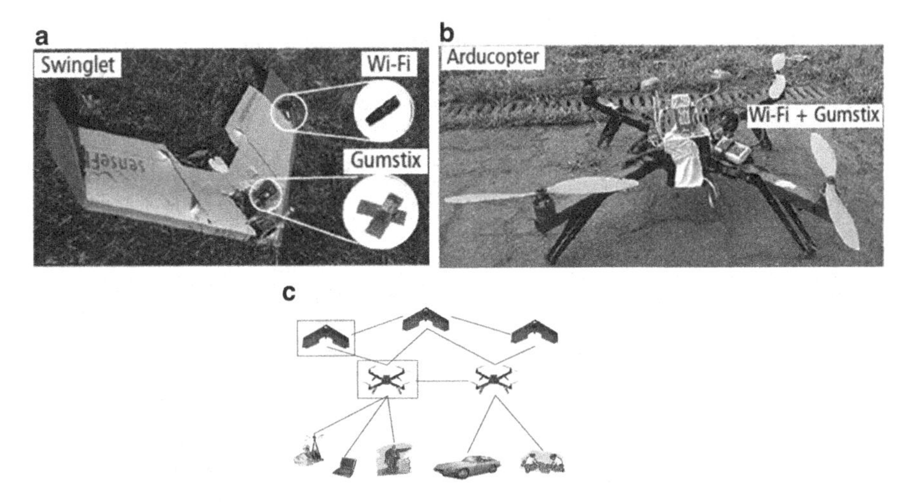

Fig. 8 A 2-layer drone network [34]. (**a**) Swinglet. (**b**) Arducopter. (**c**) Aerial network

distributed systems, sensor networks, and health areas, the principles, the utility of Big Data analytics in night vision systems have much promise for search and rescue.

On January 12, 2010, a 7.0 magnitude earthquake rocked Haiti with an epicenter that was 25 km west of Haiti's [37]. By 24 January, another 52 aftershocks with magnitude 4.5 or greater had been reported. According to incomplete statistics, more than 220,000 people were killed, 300,000 people were injured, and 1.5 million people were displaced in the disaster. Population movement, in reality, can contribute to increase morbidity and mortality and precipitate epidemics of communicable diseases in both displaced and host communities. To track and estimate population movements, Camara [36] conducted a prompt geospatial study using mobile data, as shown in Fig. 9. The mobile data was the position data of active mobile users with valid subscriber identity modules (SIM). For each SIM, a list of locations on each day during the study periods was recorded and managed in a database. The mobile and mobility data was then used to estimate the population movements and identify areas outside the city at risk of cholera outbreaks as a result of the population movements. One drawback of the use of mobile data for disaster and relief operations is the availability and fidelity of mobile data. If, for example, the mobile cellular network is down in the disaster affected areas, no mobile data can be collected. Under some scenarios, survivors that can be rescued may be, hidden under, stuck, or trapped by objects or obstacles, who are not capable of using mobile devices. This problem can be further complicated due to the existence of several population groups including the elderly, children, sick, disabled people, and pregnant woman, which RASEN night vision system could triage in advance.

The study [39] provided a review of the use of big data to aid the identification, mapping, and impact of climate change hotspots for risk communication and

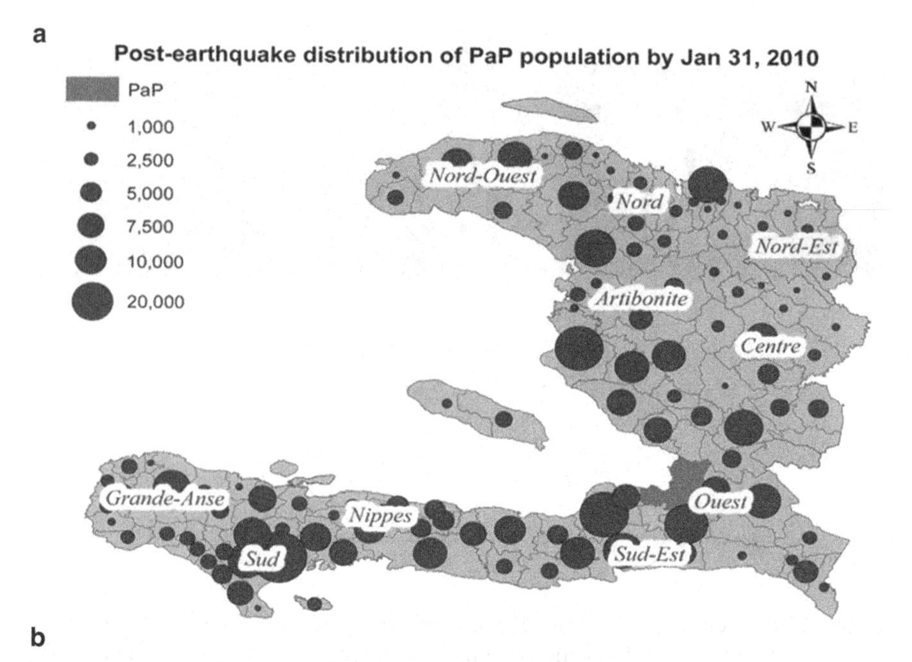

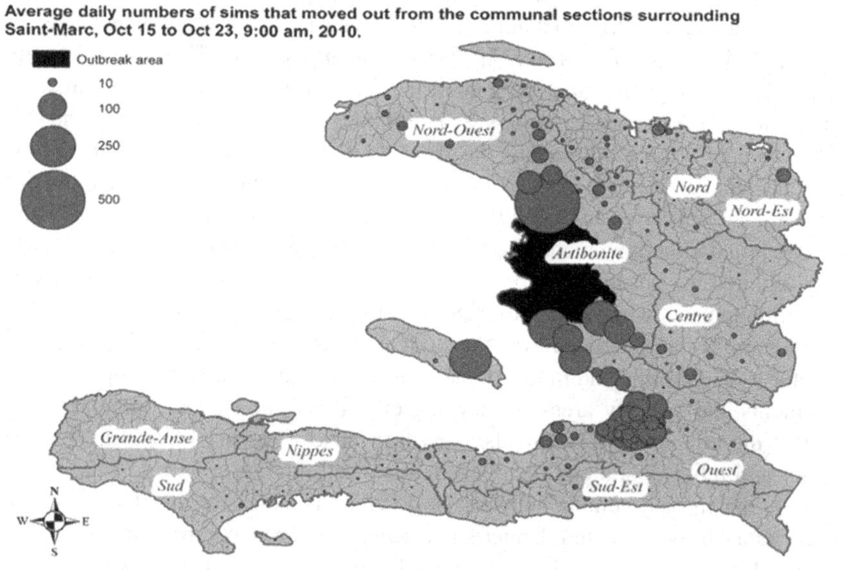

Fig. 9 Population distribution [36]. (**a**) Jan 31, 2010. (**b**) Oct 23, 2010

decision making. de Sherbinin [40] argued the data fusion for predication of the location of surface water cover and exploited the idea of bagged decision tree to derive inundation maps by combining coarse-scale remotely sensed data and fine-

Fig. 10 Segments of spatial video data [38]. (**a**) Slight damage. (**b**) Severe damage

scale topography data. It is widely recognized that the imagery is key to provide rapid, reliable damage assessments and enable quick response and rescue [38, 41]. CNN [38] employed a spatial video system to collect data by following the path of the Tuscaloosa tornado of April 27, 2011. Example segments of spatial video data are shown in Fig. 10. The spatial video data is then loaded into a GIS system ArcMap and processed offline to support post-disaster recovery. Fluet-Chouinard et al. [41] conducted the spatial video data collection four days after the tornado of April 3, 2012 in Dallas Fort-Worth (DFW) area. An online survey was then performed with the data collection to refine the damage classification, which can be referenced by further studies.

ADDSEN [42] was proposed for adaptive real-time data processing and dissemination in drone swarms executing urban sensing tasks, as shown in Fig. 11a. Two swarms of drones were dispatched and performed a distributed sensing mission. Each drone was responsible for sensing a partial area of the roadway along flight path. Instead of immediately transmitting the sensed data back to the ground control center, ADDSEN allows each drone to enable partial ordered knowledge sharing via inter-drone communication as described in Fig. 11b. Considering the drones are limited in flight time and data payload capacity, ADDSEN designed a load balancing method. In each swarm, the drone with most residual energy was selected as a balancer. The balancer relocated the work load for overload or heavy load drones and can coordinate with the drones in the same or different swarm to achieve cooperative balanced data dissemination. To enable rapidly processing big aerial data in a time-sensitive manner, Ofli et al. [43] proposed a hybrid solution combining human computing and machine intelligence. Human annotation was needed in this method to train trace data for error minimization. On the basis of trained data, image-based machine learning classifiers were able to be developed to automate disaster damage assessment process.

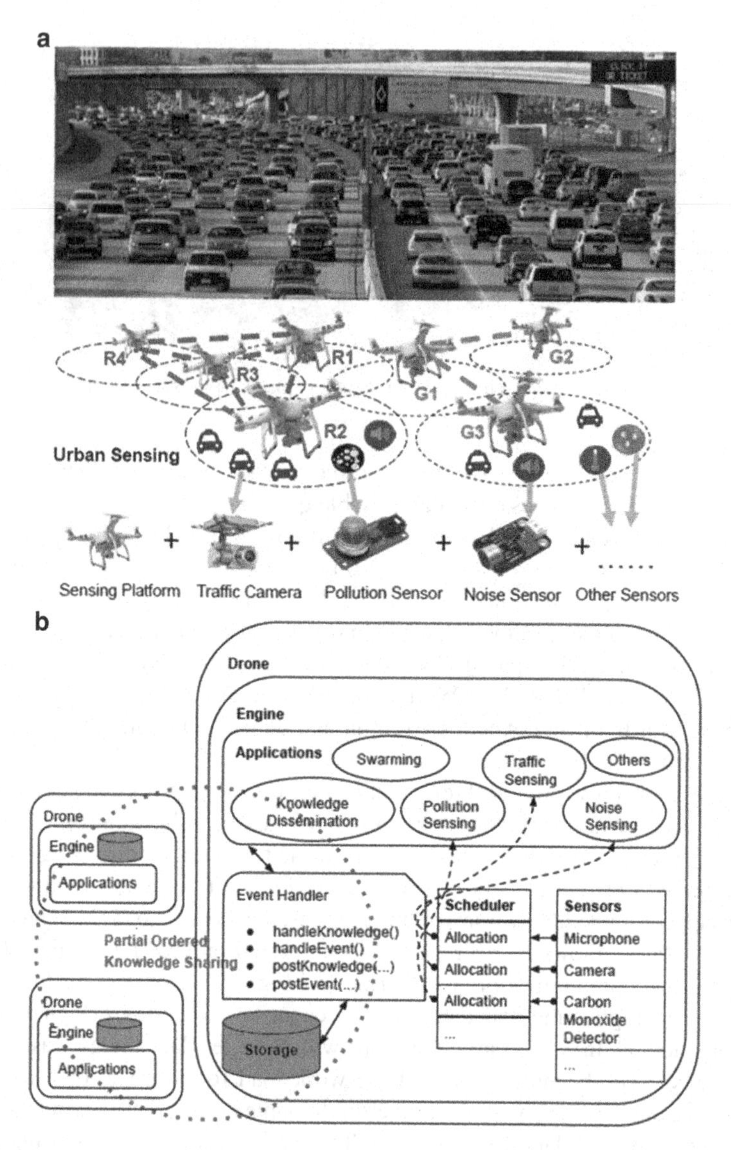

Fig. 11 Drone swarms for urban sensing [42]. (**a**) Drone swarms. (**b**) Distributed knowledge management

2.4 Swarm Intelligence

For efficient and effective search and rescue, night vision systems are required to coordinate with each other and optimize searching and sensing strategies. However, night vision systems for search and rescue exhibit complex behaviors that can be simplified and less costly when a swarm search algorithm is executed to determine a recommended path for rescue drones to traverse as first responders. As circumstances change, the swarm algorithm can adjust and recalculate with new data, providing an alternate path to rescue drones to follow for initial monitoring and injury assessment using special night vision equipment, such as RASEN, that can provide scouting details for future convoys to the disaster area. Note that the locations of objects and circumstance in target areas are often unpredictable, it is very difficult to model and analyze the behavior of night vision systems and the interactions between the systems. As matter of fact, it is desirable that night vision systems are networked and can self-organize, self-configure, accommodating to new circumstances in terms of terrain, weather, tasks, network connectivity, and visibility, etc. Our approach simplifies that adaptive response of rescue drones to such Big Data analytic environments.

Inspired by autonomy societies of insects with exact, desired characteristics, a considerable body of work on swarm intelligence (SI) for supporting rapid search and rescue has been conducted [19, 20, 43–46]. Swarm intelligence (SI) is an artificial intelligence discipline that focuses on the emergent collective behaviors of a group of self-organized individuals leveraging local knowledge. It has been observed that, as an innovative distributed intelligent paradigm, SI has exhibited remarkable efficiency in solving complex optimization problems.

The most notable examples of swarm intelligence based algorithms [47–50] are ant colony optimization (ACO), ant colony cluster optimization (ACC), boids colony optimization (BCO), particle swarm optimization (PSO), artificial bee colony (ABC), stochastic diffusion search (SDS), firefly algorithm (FA), bacteria foraging (BF), grey wolf optimizer (GWO), genetic algorithms (GA), and multi-swarm optimization (MSO).

Advanced Multi-Function Interface System (AMFIS) is an integrated system with a ground control station and a set of flight platforms developed by Fraunhofer IOSB [20]. The ground control station was deployed for controlling flight platforms and managing sensor data collection in a real-time fashion, and the flight platforms were dispatched for flight maneuvers like object tracking and data collection. Via uplink and downlink channels, the ground control station communicated with the flight platforms for controlling and transmitting data information. The uplink channel was for control while the downlink was for data transmission. It is claimed that the intelligence of the flight platforms is supported by the ground control station based on sensor data using data fusion, which is missing in [20]. Figure 12 illustrates the blueprint of AMFIS working with various sensors and mobile platforms.

RAVEN [44] is an early system that enables investigating a multi-task system with ground vehicles and drones, as shown in Fig. 13. The ground vehicles and drones were used to emulate air and ground cooperative mission scenarios. Similar to AMFIS, the coordination of ground vehicles and drones and the swarm logic are supported and provided by a central ground center station.

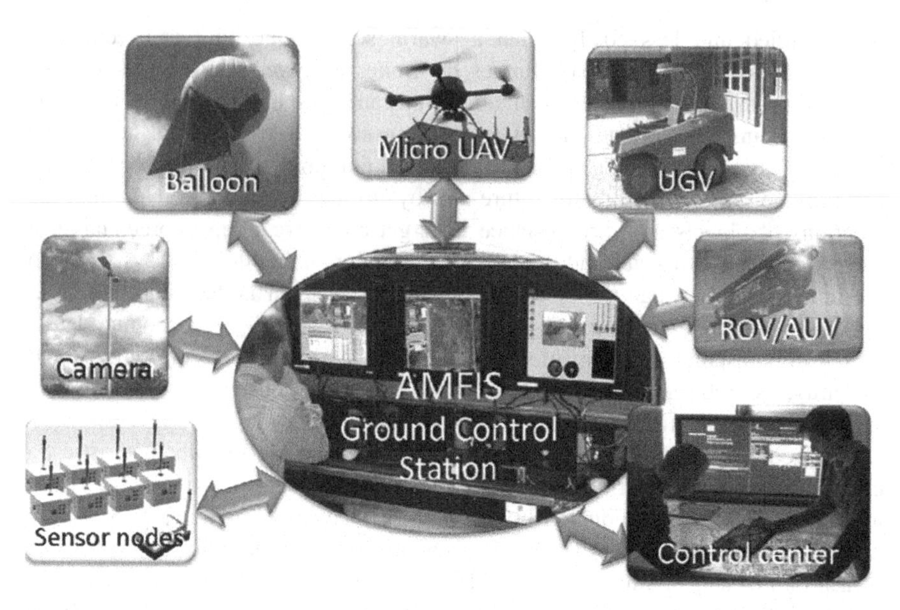

Fig. 12 AMFIS [20]

Fig. 13 RAVEN [44]

Unlike aforementioned systems, a layered dual-swarm system [46] was proposed with more detailed insight in swarm intelligences. The core of this project was focused on the intra-swarm and inter-swam intelligence in a network of wire- less sensors and mobile objects. As shown in Fig. 14, two swarm collectives were

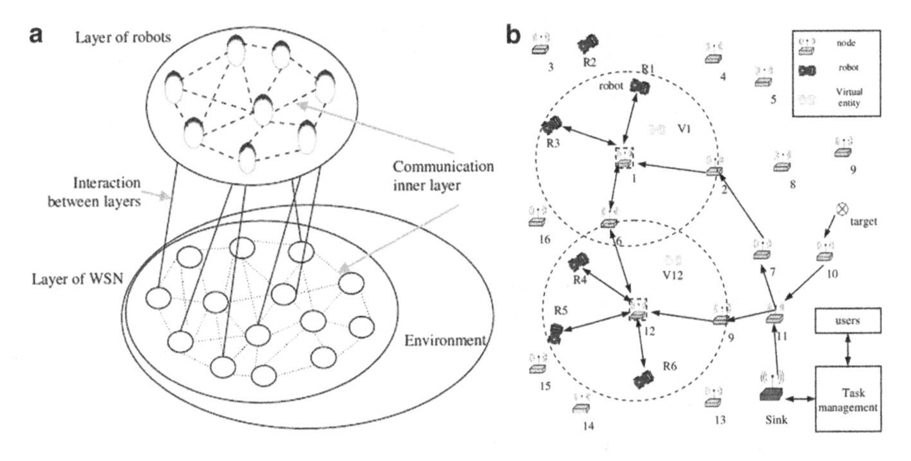

Fig. 14 A layered dual-swarm system [46]. (**a**) Layered structure. (**b**) System diagram

coexisting in a system. The upper layer consisted of autonomous mobile objects and used a boids model to guide object movements and actions.

The lower layer is a self-organized wireless sensor network, while an algorithm of ant colony swarm was applied for environmental sensing. Via a communication channel, two swarm collectives exchanged necessary information to foster cooperation between two collectives so as to form new swarm intelligence. The study [45] aimed at providing autonomous control of multiple drones. To achieve it, a function named kinematic field was introduced, which enables the drones to calculate kinematic fields on the basis of local information and autonomously plan their routes while the field was being asymmetrically modified by co-existing drones.

2.5 *Night Vision Systems*

Night vision is an ability of seeing in darkness, low illumination or night conditions. However, humans have poor night vision since there are tiny bits of visible light present and the sensitivity of human eye is quite low in such conditions. To improve visibility at night, a number of night vision devices, systems, and projects have been designed, developed, and conducted in areas. The night vision devices (NVD), also known as night optical/observation device (NOD), denote the electronically enhanced optical devices such as night vision goggles, monocular, binocular, scopes, and clip-on systems from the night vision manufacturers like Armasight Night Vision, ATN Night Vision, Yukon Night Vision, Bushnell Night Vision and others. NVDs were first used in World War II and now are available to the military, polices, law enforcement agencies, and civilian users.

Based on technology used, NVDs primarily operate in three modes: image enhancement, thermal imaging, and active illumination.

- *Image enhancement*, also called low light imaging or light amplification, collects and magnify the available light that is reflected from objects to the point that we can easily observe the image. Most consumer night vision products are image intensification devices [51]. Light amplification is less expensive than thermal imaging.
- *Thermal imaging (infrared)* operates by collecting the heat waves from hot objects that emit infrared energy as a function of their temperature such as human and animals. In general, the hotter an object is, the more radiation it emits. Thermal imaging night vision devices are widely used to detect potential security threats from great distances in low-light conditions.
- *Active illumination* works by coupling image enhancement with an active infrared illumination source for better vision. With lowest cost, active illumination night vision devices typically produce higher resolution images than that of other night vision technologies and are able to perform high-speed video capture (e.g. reading of license plates on moving vehicles). However, active illumination night vision devices that can be easily detected by other devices like night vision goggles are generally not used in tactical military operations.

Night vision devices and sensors (such as cameras, GPS, Lidar, and Radar) are integrated into night vision systems [52–56] to sense and detect objects that are difficult to see in the absence of sufficient visible light or in the blind spots. Based on the relative behavior of the night vision devices and sensors, night vision systems are commonly classified into two main categories: active and passive.

Active night vision systems equip infrared light sources and actively illuminate the objects at a significant distance ahead on the road where the headlights cannot reach. The light reflected by objects is then captured by cameras. Active systems are low cost solutions, performing well at detecting inanimate objects. In the market, automotive companies like Mercedes-Benz [56], Audi, BMW [54, 56], Rolls-Royce, GM, and Honda [57] have offered night vision systems with infrared cameras.

In the case of Audi [56, 58], BMW [56, 59], and Rolls-Royce [60, 61], Autoliv [55] systems were passive solutions. Passive systems detect thermal radiation emitted by humans, animals and other objects in the road which are processed using different filters. The object detection range can be up to 328 yards (300 m) which has twice the range of an active system and thrice the range of headlights. Honda deploys dual infrared cameras on vehicle to provide depth information for night vision. Drones may be equipped accordingly.

2.6 Artificial Cognitive Architectures

Rescue drones can also be adapted to aquatic environments considering the vast uncharted depths on Earth which is more water than land. Autonomous systems are required for navigating austere environments such as harsh landscapes on other planets and deep oceans where human analysis cannot function and directly guide drones. Hence, cognitive architectures are required and as these systems evolve and become more self-reliant and cognitive through machine learning they become increasingly valuable to search and rescue teams such as the Coast Guard and NASA. Carbone [62, 63] defines cognitive formalism within systems as a biologically inspired knowledge development workflow developed from decades of cognitive psychology research combined with neuron-like knowledge relativity threads to capture context for systems to be able to self-learn. Microsoft can store magnitudes of high volume data now in DNA and IBM continues to develop more powerful neurotropic chipsets. Artificial Cognitive Architecture research [63] is also making great strides and will provide needed improvement in levels of self-learning, context, and trust in order for autonomous systems to expand usage across difficult search and rescue environments. Therefore, it is essential to have a system that can move and think on its own, with machine learning capability, while satisfying human-driven objectives and rules optimal for SAR missions.

3 Rapid Alert System for Enhanced Night Vision (RASEN)

We developed a proposal for Sony that fused their night vision technology with emerging MMW radar that can discern detail of life behind walls from a distance, which can possibly be applied to discerning survivor status under rubble and at night as first responder rescue drones detect movement. Until very recently, efficacy of proposed approaches and systems was mostly ignored in terms of visual quality and detection accuracy with consideration of hardware cost. To help prevent accidents and increase driver awareness in a dark environment, low-cost, high accuracy, real time night vision is needed that integrates seamlessly with other smart sensors. We argue that it is essential to redesign the current architecture of night vision systems with networked vehicles and drones.

Contrary to existing architectures which rely only on drones, infrared cameras, LiDAR, or other on-board units, we develop the concept of providing capability of network-wide sensor data fusion for dynamically changing environments, particularly coupled with real-time map, weather, and traffic updates. Meanwhile, an important property of the system architecture is that it is evolvable, in the sense that it can allow far more devices or sensors mounted on vehicles, new protocols, features, and capabilities to be added on on-board platforms or the system infrastructure. The system architecture is a modularization of on-board platforms, networks, servers and technologies in which certain components (e.g., platforms and networks) remain stable, while others (the devices, sensors, links, and technologies) are encouraged to vary over time.

Figure 15 illustrates the generic framework architecture of RASEN deployment. It consists of three main components: data center and its servers, available high-speed networks including vehicular network, LTE and 5G networks, and on-board embedded platforms with the radar-camera module, sensors, and single or multiple drones. Data center and its servers are set up to achieve data and provide services to vehicles equipped with on-board platform and its modules. We aim to yield network-

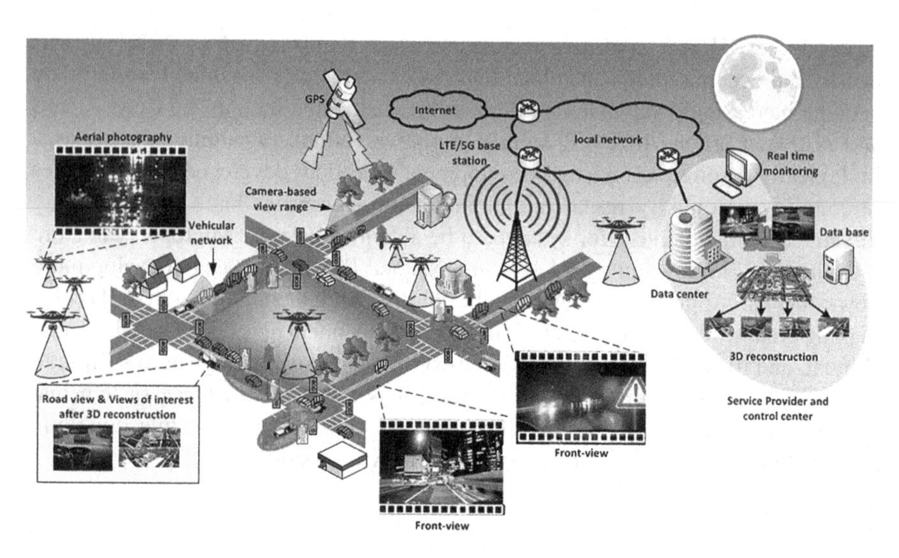

Fig. 15 Generic framework architecture of RASEN deployment

side insights on environment changes, traffic status, map updates, and weather conditions. The servers inform vehicles about real-time lightweight locational based information, which enables vehicles to know about what is ahead now on the road so that the drivers could become confident and pro-actively react to different situations. Leveraging links among servers, vehicles, and drones (i.e., remote sensing enabler) to provide network-wide machine learning capability, users (i.e., drivers) could gain the most personalized and accurate route guidance experience. Users can customize theirs interests of data from vehicle, sensors, environment, and people, etc.

Our on-board platform, as shown in Fig. 16, has an embedded PC with a TFLOP/s 256-core with NVIDIA Maxwell Architecture graphics processor unit (GPU), connecting to a 360° MMW radar, a camera system of four cameras with IMX224MQV CMOS image sensors, a Raspberry Pi3, a GPS navigation system, a DSRC module, a remote sensing system of multiple-drone.

However, due to inherent shortcoming associated with the wide signal beam, MMW radar is insensitive to the contour, feature, size, and color of the detected targets. To this end, we use high sensitivity CMOS image sensor (IMX224MQV), which is capable of capturing high-resolution color images under 0.005 lux light conditions which are nearly dark nights. On dark nights, traditional cameras typically experience low sensitivity and difficulties in discerning one color from another [51–58]. The remote sensing system is a self-organized, distributed multiple drone system to improve information fusion and situational awareness. Drones extend the limited sensing range of cameras, Lidar, MMW radar, and DSRC and provide multiple views that describe distinct perspectives of the same area circumscribing the vehicle. In some extreme circumstances, drones can serve as temporary network access points for emergency communications.

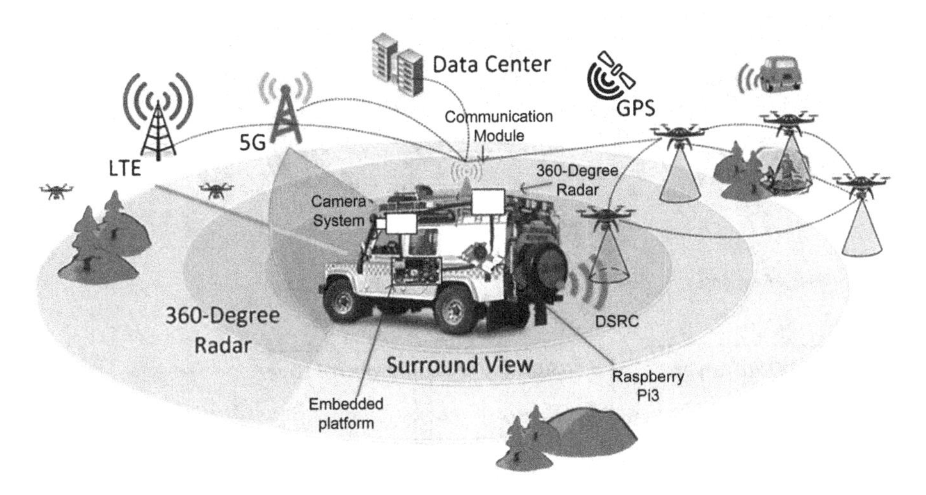

Fig. 16 On board platform configuration

By collecting and analyzing sensing data, we construct 3-layer RASEN system architecture aiming at serving various applications with demands of high accurate environmental perception like night vision, as shown in Fig. 17. Data layer collects, achieves, and unifies data representations; Fusion layer consumes the data provided by the data layer, abstracts features, detects, classifies and tracks objects; and the control layer mainly focuses on modeling situation and driving, sends, alters, and takes in-time vehicle control with respect to the information abstracted and discerned from sensing data.

In RASEN, with a geometrical model of four systems, the calibrated 360° MMW radar system, camera system, GPS navigation system, multi-drone system, and Raspberry PI work together to generate a sequence of sensing data containing environment objects through iterations of two phases. One phase explores the ability of the long detection of the multi-drone system and 360° MMW radar system. When the vehicle is moving, those systems find possible targets. Based on the remote sensing data and network-wide insights, when any target enters the vision range of the night camera system, Raspberry PI/Wolfram, a cyber-physical system platform with the capability of milliseconds processing, issues a notification message and triggers the other phase. Then the night camera starts capturing a series of low-light images. It can provide lateral resolution to analyze data and ascertain further actionable intelligence for automated vehicle systems when combined with data provided in advance by the recently developed MMW radar.

The fusion layer deals with how to fuse sensors measurements to accurately detect and consistently track neighboring objects. Each time the fusion layer receives new raw data, it reads information encoded in the data format and generates a prediction of the current set of object hypotheses [64, 65]. Features are extracted out of the measured raw data with the goal of finding all objects around the vehicle. For artifacts caused by ground detections or vegetation, we suppress them by

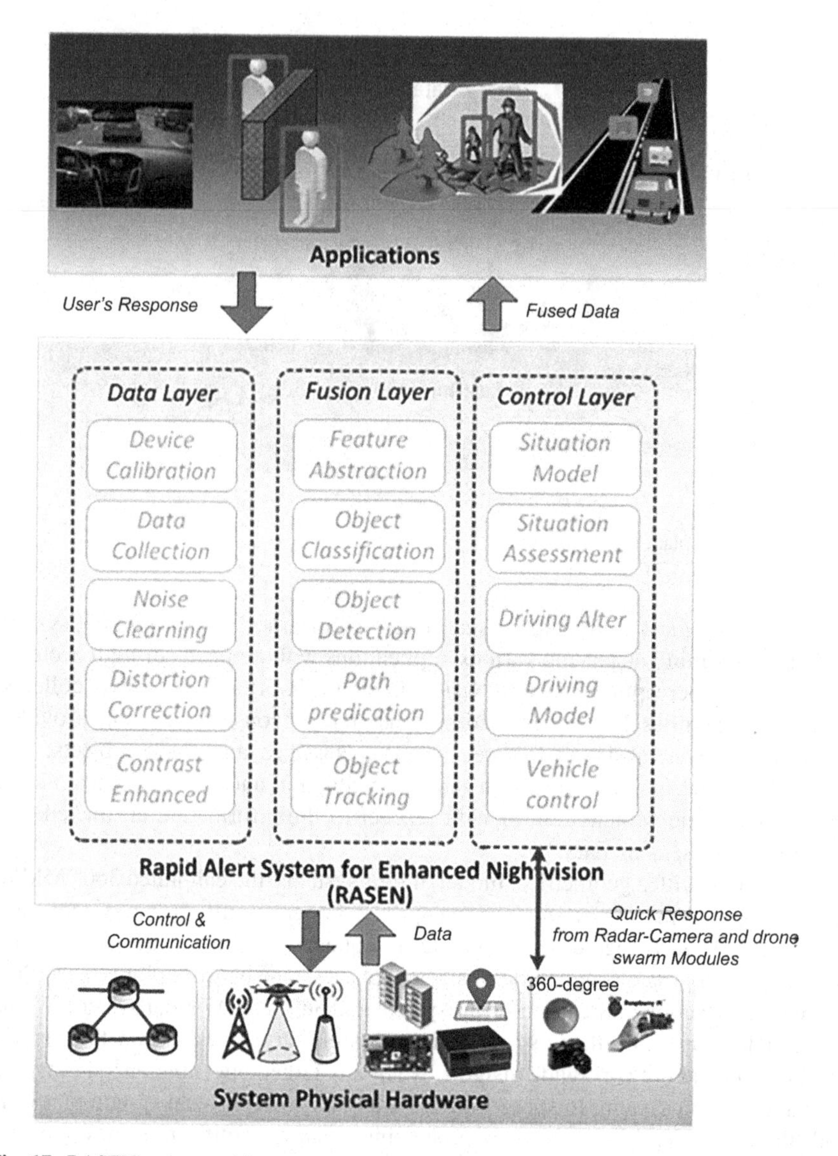

Fig. 17 RASEN system architecture

exploring their features. Both ground detections and vegetation are static and have no speed, so using the radar data, we can easily identify and mark them out. To reduce misidentification rate, 3D map information can also be used by checking against the road geometry. The result is a list of validated features that potentially originate from objects around the vehicle.

Sensing data are processed by the fusion layer and then delivered to support tasks and services in the control layer. In RASEN, network-wide information techniques are employed to assess, evaluate, and combine the information yielded from the on-board platform, in conjunction with the host-vehicle states, into reliable features which are used to improve the performance of object detection, tracking, and night vision. Besides night vision, our system is also suitable or can be extended for other applications such as smart cruise control, lane-departure warning, headlight control, active night vision, rain sensing, and road sign recognition.

4 Swarm Intelligence Utilizing Networked RFID

Modern developments in wireless technology have increased the reliability and throughput of this type of communication. The factors of portability, mobility, and accessibility have all improved. We will apply metaheuristics of Ant-Optimization using Wolfram language to a previous work of Antenna Networks [62, 63, 66]. We will review below three widely used systems, which are Radio Frequency Identification, Wireless Sensor Networks, and Multiple-Input Multiple-Output communication.

4.1 Radio Frequency Identification (RFID) for Wireless Drone Networking

Radio Frequency Identification (RFID) is a technology that uses a radio frequency electromagnetic field to identify objects through communication with tags that are attached to them. This technology originally was introduced during World War II [67]. Figure 18 depicts RFID components in context of a communication network. The RFID system consists of two components: readers and tags or namely interrogators and transponders. Each reader and tag has antenna to communicate wirelessly through electromagnetic waves (See Fig. 19). There are two types of RFID tags, which are Active and Passive. The active RFID tags contain internal power source and the passive RFID tags usually harvest energy from readers' signals.

Wireless Sensor Networks (WSNs) consist of spatially distributed autonomous devices using sensors to monitor physical or environmental conditions as shown in Fig. 19. A WSN is used in many industrial, military, and consumer applications. The WSN consists of nodes where each node is connected to a single or multiple sensors. Typically, each sensor network node has multiple components [69]:

- *Transceiver* with an internal antenna or connection for external antenna.
- *Microcontroller*, an electronic circuit, with sensor interface as well as energy source interface which is usually a battery or an embedded form of energy harvesting.

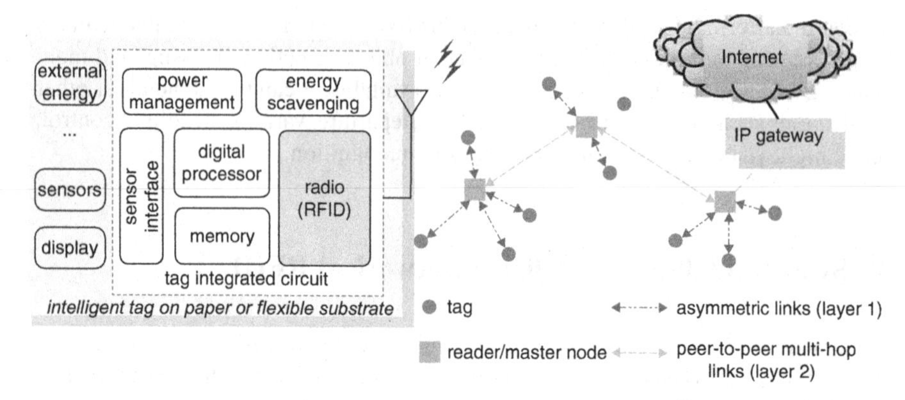

Fig. 18 A generic IoT platform consisting of intelligent RFID tag and reader with a hierarchical two-layer network [68]

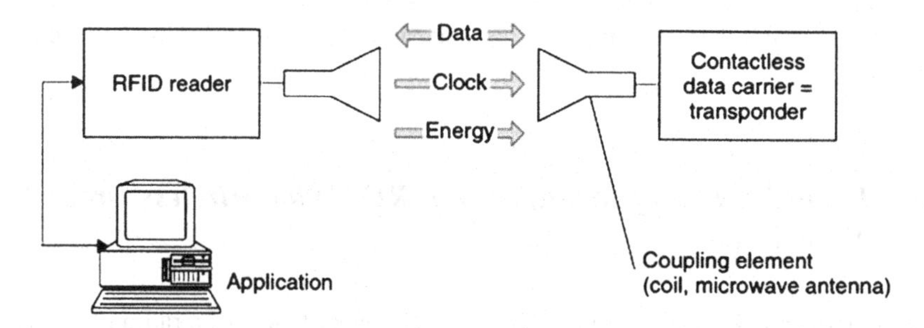

Fig. 19 A typical RFID system wireless sensor networks [67]

A sensor is a device that receives a signal or stimulus from the surrounding and responds to it in a distinctive manner. It converts any mechanical, chemical, magnetic, thermal, and electrical or radiation quantity into measurable output signal. The basic features and properties of sensors are:

- *Sensitivity*: This represents the detection capability of the sensor with respect to the sample concentration.
- *Selectivity*: This represents the ability to detect the desired quantity among other non-desired quantity.
- *Response time*: This describes the speed in which the sensor can react to changes.
- *Operating life*: This is the lifetime of the sensor.

A WSN topology can vary from a simple star network to a complex multi-hop wireless mesh network and the propagation between the hops of WSN can be routing or flooding [70] (Fig. 20).

The concepts of multiple-input multiple-output (MIMO) in wireless communication is based on the use of multiple antennas at both the source (transmitter) and the destination (receiver) to exploit multipath propagation [71]. MIMO is a developed

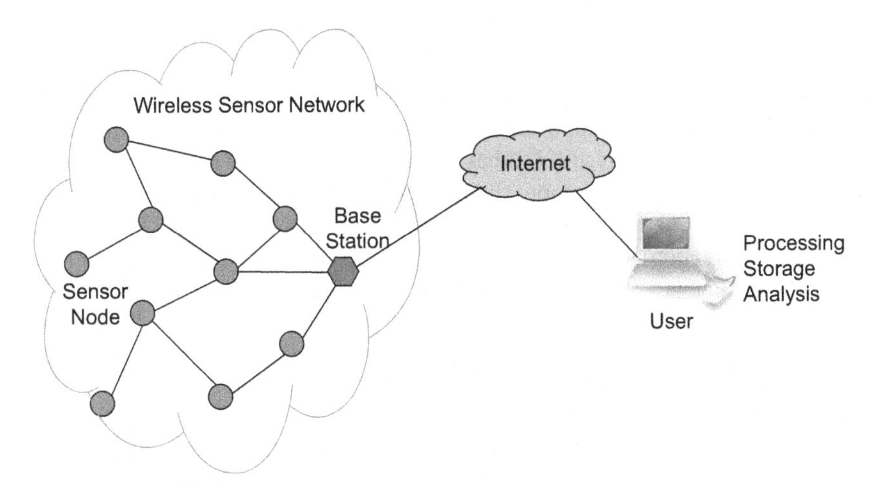

Fig. 20 A typical wireless sensor network via multiple-input multiple-output (MIMO) communication

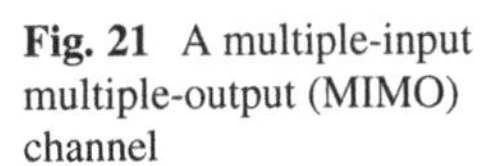
Fig. 21 A multiple-input multiple-output (MIMO) channel

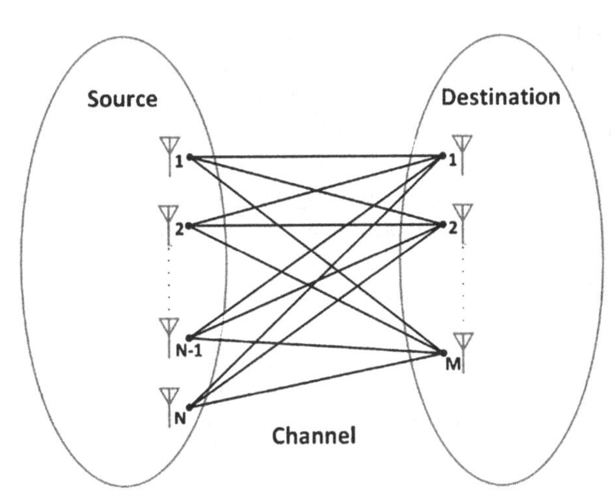

form of antenna array communication that provides advantages such as gain and spatial diversity. Although multiple receive antennas have been known and used for some time, the use of transmit diversity has only been studied recently [65]. Figure 21 illustrates the concept of MIMO system with source transmit antennas and destination receive antennas.

In modeling and analysis of antenna systems in settings of RFID systems, networks of sensors, and MIMO situations, we implemented a simulation using metaheuristics of Ant-Optimization and Particle Optimization with Wolfram language. Although the examples are limited to small number of nodes, due to the nature of the approach and its scalability, this model represents a step towards scaling to Big Data related problems in these situations.

4.2 Ant-Colony Meta-Heuristics for Night Rescue Operations

In the early 1990s, ant colony optimization (ACO) was introduced by M. Dorigo and associates as a bio-inspired metaheuristic for the solution of combinatorial optimization (CO) problems [72]. It has been stated that "ACO belongs to the class of metaheuristics which are approximate algorithms used to obtain good enough solutions to hard CO problems in a reasonable amount of computation time". We are adapting this approach to rescue drones that need to determine, in CO problems posed by disaster areas, an optimal path to survivors in a reasonable amount of time using ACO approach. It is known that the inspiration of ACO is based on the behavior of real ants. When searching for food, ants initially explore randomly surrounding environment of their nest. If an ant finds a food source, it carries a sample back to the nest. During this trip, it is known that the ant leaves a Chemical (pheromone) trail. The quantity of pheromone deposited guides other ants to the food source. This indirect communication among ants through pheromones provides a mechanism to find near-optimal paths between their nest and sources. Naturally this approach results in a swarm convergence toward the shortest trail to food sources. This colony metaheuristic approach modeling ant behavior in nature has been used successfully to find near-optimal solutions to relatively large unstructured network problems, which we are applying to rescue drones determining optimal path to survivors. Instead of deploying expensive drones to survey an expanse of disaster area to determine best way to deploy relief convoys, our approach simplifies the first responder search by deploying more, inexpensive scout drones that can immediately execute a swarm algorithm that will provide a recommended path to follow between survivors. This approach saves on parameters such as rescue time and fuel cost when surveying vast areas with drones to determine optimal path by calculating in advance using a swarm algorithm to precisely plot a course of action for the drones to execute, with capability to recalculate and adjust network as conditions change. Wireless sensor bandwidth can then be made available directly for survivor search with less costly drones, with increased payload availability for RASEN and other night vision systems, rather than deploying more sophisticated drones to survey a vast expanse of area for pinpoint accuracy.

Adapting the Ant-Colony metaheuristic as implemented in Mathematica by Rasmus Kamper, we started with a random set of antenna nodes (these antennas, depending on the problem, can belong to actual drones). As explained in the above wireless networks section, we demonstrated that the Ant-Colony algorithm would find the solution in reasonable time and iterations. The progressive iterations of the algorithm applied to the networks of antenna problem are shown in Fig. 22. The advantages of using ant-colony algorithm before actual deployment of swarm of drones are many. Firstly, the actual communication among drones would either be avoided or minimized, since a near-optimal visit patterns of the drones are identified already through the process ant-colony optimization. Therefore, cheaper and less capable drones could be used as well.

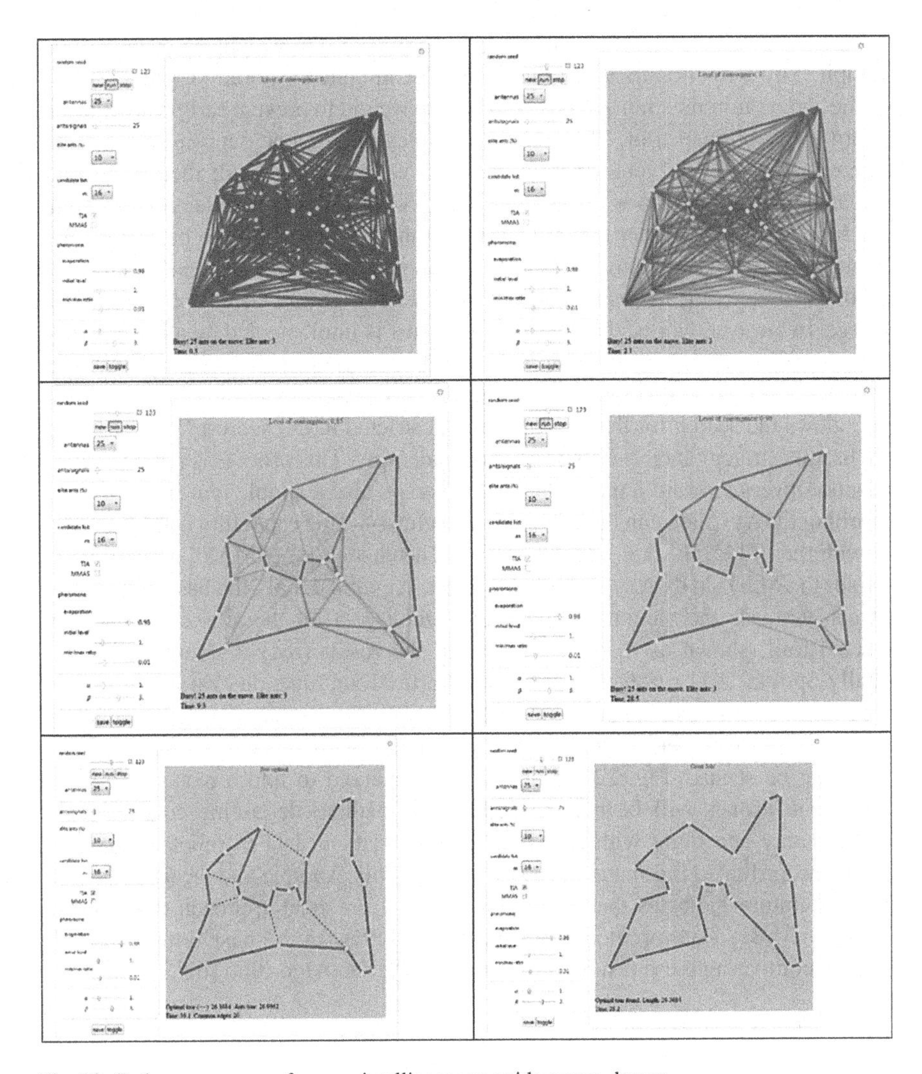

Fig. 22 Path convergence of swarm intelligence to guide rescue drones

Figure 22 illustrates the operation of the algorithm. First step is the identification of size of disaster sites (DS) (notion of survivors on location). The DS are represented by random points in a two dimensional space and the distance between each pair of DS (represented by the edge weight) is taken as the Euclidean distance. Algorithms proceed to construct tours to visit the DSs until convergence after the first initialization. Algorithm also simulates evaporation of deposited chemicals called "pheromones", which provide a basis for AI weighting. It should be noted that as ants explore the options, all ants complete Hamiltonian cycles by starting from a randomly selected DS. At the time of initialization all edges are assigned equal weights (pheromone) which can be controlled by the "initial level" slider in the visualization screen. After each construction step, the weight on each edge is multiplied by a fraction to adjust the weight to simulate evaporation. The user can use the slider "min/max ratio" to set a minimum weight to prevent early convergence toward a sub-optimal trail. The update simulates pheromone deposit by a weight increase. It should be noted that shorter trail edges are favored; thereby weight increase (pheromone deposited) is inversely proportional to trail length. An edge with higher weight (shorter length) is assigned higher probability to be selected by an ant. As this process continues, the edges on the graph with less traffic will fade as observed by color intensity, and the ants' preferred trail (near-optimized) will emerge. In large networks, the notion of elite ants is implemented by allowing only the most efficient ants to increase weights (deposit pheromone). The menu "elite ants" in the screen controls the corresponding percentage. In addition, a "candidate list" allows the search for the next DS to be restricted to the nearest DS. For larger graphs, this strategy increases the speed considerably. This process is repeated until all ants converge toward a particular trail. There is also a simple tour improvement algorithm (TIA) which can be activated on each tour. This capability can be turned-off with the "TIA" checkbox on the screen. The checkbox "MMAS" on the screen enables the MAX-MIN Ant System algorithm. If enabled, only the best-performing ant can increase the weights (deposit pheromone). After the ants converge to a selected trail, shown as read in the screen, the result (red) is compared to the usually optimal trail (dashed) using FindShortestTour function. Various runs can be performed by saving the outcome of a particular run. One can change parameters and run again to see performance on the same graph with the new settings.

As can be seen in Fig. 22 the algorithm converged to find a reasonable path. The rescue drones will be able to follow this path to determine conditions of survivors day and night with RASEN, relaying critical data to relief convoys to prepare according to the status of each DS node. This example and recent examples in the literature indicates that ACO research is a practical approach to scale for unstructured Big Data problems with visual analytics. In the near future, we will work on comparing these results with our Least Action Algorithm [62].

5 Conclusion

Swarm intelligence algorithms combined with Rapid Alert Sensor for Enhanced Night vision (RASEN) can provide continuous night search capability and survivor condition identification hidden under rubble for first responders during rescue operations. The Wolfram Framework provides an environment to for research students to expand their capability to develop smart rescue drones with decision support functions such as swarm intelligence. We have introduced biologically inspired algorithms combined with fusion night vision technology that can rapidly converge on a near optimal path in reasonable time between survivors and identify signs of life trapped in rubble. Wireless networking with dynamic programming ability to determine near optimal path using Big Data analytic visualization is provided to rescue teams utilizing drones as first responders based on the results of swarm intelligence algorithms. This automated multiple drone scout approach enables appropriate relief supplies to be deployed intelligently by networked convoys to survivors continuously throughout the night, within critical constraints calculated in advance by rescue drones, such as projected time, cost, and energy per mission.

References

1. VOA: Natural disasters cause $175 billion in damages. http://www.voanews.com/a/years-natural-disasters-cause-billions-of-dollars-in-damages/3663333.html (2016). [Online]; Accessed 21 May 2016
2. Owens, D.A., Sivak, M.: Differentiation of visibility and alcohol as contributors to twilight road fatalities. Hum. Factors. **38**(4), 680–689 (1996)
3. Federal Highway Administration: Driving after dark. https://www.fhwa. dot.gov/publications/publicroads/03jan/05.cfm (2016). [Online]; Accessed 10 Mar 2016
4. Nathaniel F.: EO soldiers new integrated rifle/helmet night vision system. http://www.thefirearmblog.com/blog/2015/07/28/peo-soldiers-new-integrated-riflehelmet-night-vision-system/ (2016). [Online]; Accessed 29 July 2016
5. FLIR: Signal detection & ITS video. http://www.flir.com/security/display/?id=44284 (2016). [Online]; Accessed 15 May 2016
6. WIKI: Automotive night vision. https://en.wikipedia.org/wiki/Automotive_night_vision (2016). [Online]; Accessed 17 Dec 2016
7. ROBOHUB: Swarms of precision agriculture robots could help put food on the table. http://robohub.org/swarms-of-precision-agriculture-robots-could-help-put-food-on-the-table/ (2016). [Online]; Accessed 21 Oct 2016
8. Interesting Engineering: Swedens futuristic fire-fighting helmet. http://interestingengineering.com/swedens-futuristic-fire-fighting-helmet/ (2016). [Online]; Accessed 15 Mar 2016
9. Security Systems Technology: Thermal imaging technology for surveillance and security. http://www.securitysystems-tech.com/files/securitysystems/supplier_docs/ULIS-Surveillance%20and%20Security.pdf (2016). [Online]; Accessed 18 June 2016
10. Waharte, S., Trigoni, N.: Supporting search and rescue operations with UAVs. In: 2010 International Conference on Emerging Security Technologies (EST), pp. 142–147. IEEE (2010)
11. Asadpour, M., Giustiniano, D., Anna Hummel, K., Egli, S.: UAV networks in rescue missions. In: Proceedings of the 8th ACM International Workshop on Wireless Network Testbeds, Experimental Evaluation & Characterization, pp. 91–92. ACM (2013)
12. Hsu, H.J., Chen, K.T.: Face recognition on drones: issues and limitations. In: Proceedings of the First Workshop on Micro Aerial Vehicle Networks, Systems, and Applications for Civilian Use, pp. 39–44. ACM (2015)
13. Wallace, L., Lucieer, A., Watson, C., Turner, D.: Development of a UAV-LiDAR system with application to forest inventory. Remote Sens. **4**(6), 1519–1543 (2012)
14. Lin, L., Goodrich, M.A.: UAV intelligent path planning for wilderness search and rescue. In: IEEE/RSJ International Conference on Intelligent Robots and Systems, 2009. IROS 2009, pp. 709–714. IEEE (2009)
15. Almurib, H.A.F., Nathan, P.T., Kumar, T.N.: Control and path planning of quadrotor aerial vehicles for search and rescue. In: 2011 Proceedings of SICE Annual Conference (SICE), pp. 700–705. IEEE (2011)
16. Gade, R., Moeslund, T.B.: Thermal cameras and applications: a survey. Mach. Vis. Appl. **25**(1), 245–262 (2014)

17. Dong, N., Ren, X., Sun, M., Jiang, C., Zheng, H.: Fast stereo aerial image construction and measurement for emergency rescue. In: 2013 Fifth International Conference on Geo-Information Technologies for Natural Disaster Management (GiT4NDM), pp. 119–123. IEEE (2013)
18. Morse, B.S., Engh, C.H., Goodrich, M.A.: UAV video coverage quality maps and prioritized indexing for wilderness search and rescue. In: 2010 5th ACM/IEEE International Conference on Human-Robot Interaction (HRI), pp. 227–234. IEEE (2010)
19. Natraj, A., Sturm, P., Demonceaux, C., Vasseur, P.: A geometrical approach for vision based attitude and altitude estimation for UAVs in dark environments. In: 2012 IEEE/RSJ International Conference on Intelligent Robots and Systems (IROS), pp. 4565–4570. IEEE (2012)
20. Burkle, A., Segor, F., Kollmann, M.: Towards autonomous micro UAV swarms. J. Intell. Robot. Syst. **61**(1–4), 339–353 (2011)
21. Gharibi, M., Boutaba, R., Waslander, S.L.: Internet of drones. IEEE Access. **4**, 1148–1162 (2016)
22. Kothari, M., Postlethwaite, I.: A probabilistically robust path planning algorithm for UAVs using rapidly-exploring random trees. J. Intell. Robot. Syst. **71**(2), 231–253 (2013)
23. Sujit, P.B., Saripalli, S., Sousa, J.B.: Unmanned aerial vehicle path following: a survey and analysis of algorithms for fixed-wing unmanned aerial vehicles. IEEE Control Syst. **34**(1), 42–59 (2014)
24. Wikipedia: General Atomics MQ-9 Reaper. https://en.wikipedia.org/wiki/ [Online]; Accessed 21 May 2016
25. Daniel, K., Dusza, B., Lewandowski, A., Wietfeld, C.: Air-shield: a system-of-systems MUAV remote sensing architecture for disaster response. In: Systems Conference, 2009 3rd Annual IEEE, pp. 196–200. IEEE (2009)
26. Remy, G., Senouci, S.-M., Jan, F., Gourhant, Y.: SAR drones: drones for advanced search and rescue missions. Journées Nationales des Communications dans les Transports. **1**, 1–3 (2013)
27. Hammerseth, V.B.: Autonomous unmanned aerial vehicle in search and rescue - a prestudy. https://daim.idi.ntnu.no/masteroppgaver/010/10095/masteroppgave.pdf/(2017). [Online]; Accessed 14 Oct 2017
28. Liu, Z., Li, Z., Liu, B., Fu, X., Raptis, I., Ren, K.: Rise of mini-drones: applications and issues. In: Proceedings of the 2015 Workshop on Privacy-Aware Mobile Computing, pp. 7–12. ACM (2015)
29. PhantomPilots: Airnamics, what a big drone you have!! http://www.phantompilots.com/threads/airnamics-what-a-big-drone-you-have.64624/ (2017) [Online]; Accessed 1 Feb 2017
30. Kim, G.H., Nam, J.C., Mahmud, I., Cho, Y.Z.: Multi-drone control and network self-recovery for flying ad hoc networks. In: 2016 Eighth International Conference on Ubiquitous and Future Networks (ICUFN), pp. 148–150. IEEE (2016)
31. Tanzi, T.J., Chandra, M., Isnard, J., Camara, D., Sebastien, O., Harivelo, F.: Towards "drone-borne" disaster management: future application scenarios. ISPRS Annals of Photogrammetry, Remote Sensing and Spatial Information Sciences. **3**(8), 181–189 (2016)
32. Cimino, M.G.C.A., Celandroni, N., Ferro, E., La Rosa, D., Palumbo, F., Vaglini, G.: Wireless communication, identification and sensing technologies enabling integrated logistics: a study in the harbor environment. CoRR. abs/1510.06175(10). arXiv preprint (2015)
33. Rahman, M.A.: Enabling drone communications with wimax technology. In: The 5th International Conference on Information, Intelligence, Systems and Applications, IISA 2014, pp. 323–328. IEEE (2014)
34. Bekmezci, I., Sahingoz, O.K., Temel, S.: Flying ad-hoc networks (FANETs): a survey. Ad Hoc Netw. **11**(3), 1254–1270 (2013)
35. Asadpour, M., Van den Bergh, B., Giustiniano, D., Hummel, K., Pollin, S., Plattner, B.: Micro aerial vehicle networks: an experimental analysis of challenges and opportunities. IEEE Commun. Mag. **52**(7), 141–149 (2014)

36. Camara, D.: Cavalry to the rescue: drones fleet to help rescuers operations over disasters scenarios. In: 2014 IEEE Conference on Antenna Measurements & Applications (CAMA), pp. 1–4. IEEE (2014)
37. Bengtsson, L., Lu, X., Thorson, A., Garfield, R., Von Schreeb, J.: Improved response to disasters and outbreaks by tracking population movements with mobile phone network data: a post-earthquake geospatial study in Haiti. PLoS Med. **8**(8), e1001083 (2011)
38. CNN: Haiti earthquake fast facts. http://www.cnn.com/2013/12/12/world/haiti-earthquake-fast-facts/ (2013). [Online]; Accessed 21 Dec 2013
39. Curtis, A., Mills, J.W.: Spatial video data collection in a post-disaster landscape: the Tuscaloosa tornado of April 27th 2011. Appl. Geogr. **32**(2), 393–400 (2012)
40. de Sherbinin, A.: Climate change hotspots mapping: what have we learned? Clim. Change. **123**(1), 23–37 (2014)
41. Fluet-Chouinard, E., Lehner, B., Rebelo, L.-M., Papa, F., Hamilton, S.K.: Development of a global inundation map at high spatial resolution from topographic downscaling of coarse-scale remote sensing data. Remote Sens. Environ. **158**, 348–361 (2015)
42. Barrington, L., Ghosh, S., Greene, M., Har-Noy, S., Berger, J., Gill, S., Lin, A.Y.-M., Huyck, C.: Crowdsourcing earthquake damage assessment using remote sensing imagery. Ann. Geophys. **54**(6), (2012)
43. Ofli, F., Meier, P., Imran, M., Castillo, C., Tuia, D., Rey, N., Briant, J., Millet, P., Reinhard, F., Parkan, M., et al.: Combining human computing and machine learning to make sense of big (aerial) data for disaster response. Big Data. **4**(1), 47–59 (2016)
44. Camerono, S., Hailes, S., Julier, S., McCleanu, S., Parru, G., Trigonio, N., Ahmed, M., McPhillips, G., De Nardil, R., Nieu, J., et al.: SUAAVE: combining aerial robots and wireless networking. In: 25th Bristol International UAV Systems Conference, pp. 1–14. Citeseer, Princeton, NJ (2010)
45. How, J.P., Behihke, B., Frank, A., Dale, D., Vian, J.: Real-time indoor autonomous vehicle test environment. IEEE Control Syst. **28**(2), 51–64 (2008)
46. Clark, R., Punzo, G., Dobie, G., Summan, R., MacLeod, C.N., Pierce, G., Macdonald, M.: Autonomous swarm testbed with multiple quadcopters. In: 1st World Congress on Unmanned Systems Enginenering, 2014-WCUSEng (2014)
47. Li, W., Shen, W.: Swarm behavior control of mobile multi-robots with wireless sensor networks. J. Netw. Comput. Appl. **34**(4), 1398–1407 (2011)
48. Engelbrecht, A.P.: Fundamentals of Computational Swarm Intelligence. Wiley, Chichester (2006)
49. Kolias, C., Kambourakis, G., Maragoudakis, M.: Swarm intelligence in intrusion detection: a survey. Comput. Secur. **30**(8), 625–642 (2011)
50. Ab Wahab, M.N., Nefti-Meziani, S., Atyabi, A.: A comprehensive review of swarm optimization algorithms. PloS One. **10**(5), e0122827 (2015)
51. Chu, S.C., Huang, H.C., Roddick, J.F., Pan, J.S.: Overview of algorithms for swarm intelligence. In: International Conference on Computational Collective Intelligence, pp. 28–41. Springer (2011)
52. American Technology Network Corp: Night vision technology. https://www.atncorp.com/hownightvisionworks (2016). [Online]; Accessed 16 Mar 2016
53. CVEL: Night vision systems. http://www.cvel.clemson.edu/auto/systems/night-vision.html (2016). [Online]; Accessed 12 Dec 2016
54. Josh Briggs: How in-dash night-vision systems work. http://electronics.howstuffworks.com/gadgets/automotive/in-dash-night-vision-system.htm (2016). [Online]; Accessed 30 Aug 2016
55. Computer Reports: Driving a BMW with night vision proves illuminating. http://www.consumerreports.org/cro/-news/2014/10/driving-a-bmw-with-night-vision-proves-illuminating/index.htm (2016). [Online]; Accessed 18 Sept 2016
56. Autoliv AB: Autolive. https://www.autoliv.com/ (2016) [Online]; Accessed 6 Mar 2016
57. Pund, D.: Night light: night-vision systems compared from BMW, Mercedes, and Audi. http://www.caranddrive.com/com-parisons/ night-vision-systems-compared-bmw-vs-mercedes-benz-vs-audi-comparison-test/. (2016). [Online]; Accessed 21 May 2016

58. Austin, M.: Intelligent adaptive cruise, autonomous braking, and green wave: what Honda's got up its sleeve. https://www.cnet.com/roadshow/news/autoliv-aims-night-vision-at-mass-market/ (2016). [Online]; Accessed 21 Nov 2016
59. Autoliv: Audi night vision assistant. http://www.autolivnightvision.com/vehicles/audi/ (2016) [Online]; Accessed 8 Nov 2016
60. Ahire, A.S.: Night vision system in BMW. International Review of Applied Engineering Research. **4**(1), 1–10 (2014)
61. Autoliv: Night vision for Rolls-Royce. http://autoliv2.managecontent.com/the-technology-behind-night-vision/rolls-royce (2016). [Online]; Accessed 15 Nov 2016
62. Road Show: Autoliv aims night vision at mass market. https://www.cnet.com/roadshow/news/autoliv-aims-night-vision-at-mass-market/ (2016). [Online]; Accessed 15 Nov 2016
63. Crowder, J.A., Carbone, J.N., Friess, S.A.: Artificial Cognition Architectures. Springer, New York (2014)
64. Carbone, J.N.: A Framework for enhancing transdisciplinary research knowledge. PhD dissertation, Texas Tech University (2010)
65. Urmson, C., Anhalt, J., Bagnell, D., Baker, C., Bittner, R., Clark, M., Dolan, J., Duggins, D., Galatali, T., Geyer, C., et al.: Autonomous driving in urban environments: boss and the urban challenge. J. Field Rob. **25**(8), 425–466 (2008)
66. Xu, F., Liu, X., Fujimura, K.: Pedestrian detection and tracking with night vision. IEEE Trans. Intell. Transp. Syst. **6**(1), 63–71 (2005)
67. Bliss, D.W., Forsythe, K.W., Chan, A.M.: MIMO wireless communication. Lincoln Lab. J. **15**(1), 97–126 (2005)
68. Shannon, C.E.: Communication in the presence of noise. Proc. IRE. **37**(1), 10–21 (1949)
69. Berge, C.: Stability number. In: Graphs and Hypergraphs, pp. 272–302. American Elsevier Publishing Company, New York (1973)
70. Dargie, W., Poellabauer, C.: Fundamentals of Wireless Sensor Networks: Theory and Practice. Wiley, Chichester (2010)
71. Güldal, S.: Modeling and analysis of systems using the least action principle in information theory. University of Alabama at Birmingham, Interdisciplinary Engineering Program. Birmingham, AL: Ph.D. dissertation, Interdisciplinary Engineering Program, University of Alabama at Birmingham (2016)
72. Dorigo, M., Blum, C.: Ant colony optimization theory: a survey. Theor. Comput. Sci. **344**(2–3), 243–278 (2005)
73. Wu, D., Arkhipov, D.I., Kim, M., Talcott, C.L., Regan, A.C., McCann, J.A., Venkatasubramanian, N.: Addsen: adaptive data processing and dissemination for drone swarms in urban sensing. IEEE Trans. Comput. **66**(2), 183–198 (2017)

Gender Classification Based on Deep Learning

Dhiraj Gharana, Sang C. Suh, and Mingon Kang

1 Introduction

Gender classification from facial images is a fundamental research component that plays important roles in a wide range of real-world applications such as human-computer interaction, surveillance cameras, mobile security, and smart digital signage. The increasing business demands for gender classification from digital facial images and videos have triggered active research in the fields. Recently, image classification has been remarkably advanced by deep learning using convolutional neural networks. Especially, convolutional neural networks have been shifting the paradigm for gender classification from facial images by dramatically improving the performance.

Convolutional neural networks perform hierarchical feature extraction. The first convolutional layer automatically extracts relevant features from the facial images (e.g., lines or curves of various patterns). Then, the following convolutional layers build higher abstractions from the outputs of the (adjacent) previous layers. In gender classification, a convolutional neural network builds feature abstractions based on the differences between males and females with regard to parts of a face such as hair, eyebrows, lips, and chin.

However, the performance in the task of gender classification and the computational cost for training deep learning neural networks depend on neural network architectural settings. The settings in neural networks include the number of layers, the number of filters and the size of filters in the convolutional layer, the number of

D. Gharana • M. Kang (✉)
Department of Computer Science, Kennesaw State University, Marietta, GA, USA
e-mail: mkang9@kennesaw.edu

S.C. Suh
Department of Computer Science, Texas A&M University-Commerce, Commerce, TX, USA

S.C. Suh, T. Anthony (eds.), *Big Data and Visual Analytics*,
https://doi.org/10.1007/978-3-319-63917-8_3

neurons in the fully-connected layers, and the activation and cost functions. Most deep learning-based methods define the neural network architecture empirically.

In this chapter, we aim to (1) introduce a conventional deep learning approach to gender classification from facial images and (2) to provide a quantitative analysis of the performance for various architectures of the conventional deep learning approach with respect to the prediction accuracy and the training time. First, we present the background of deep learning to readers who are interested in deep learning for gender classification. It would help them understand neural networks, convolutional neural networks, and deep learning structures that many deep learning studies have conventionally considered for computer vision problems. Furthermore, we also provide an insight into the optimal structure of the conventional convolutional neural networks for gender classification based on our experimental results. We conducted intensive experiments by varying the number of convolutional and fully-connected layers. Most deep learning structures have been empirically constructed, and there have seldom been research works that compare the performances of various deep learning structures. Thus, the comparative experiment would provide an intuition to define a deep learning architecture.

This chapter is organized as follows. In Sect. 2, we introduce the related literature of preprocessing techniques, non-deep learning and deep learning-based gender classification methods. In Sect. 3, we briefly describe deep learning structures for gender classification. In Sect. 4, we introduce the datasets that we have used for training and test. In Sect. 5, we elucidate experimental settings and results for the optimal deep learning architecture. Then, we finalize this chapter with the discussion in Sect. 6.

2 Related Works

A number of automated gender classification studies have been conducted due to its importance in many research domains and the increasing business demands. Much research has shed light on the use of various data preprocessing techniques, filtering techniques, and machine learning tools for gender classification.

A number of image preprocessing techniques have been proposed to enhance performance for gender classification [1–3]. The performance in using the three different preprocessing techniques—face detection by Viola-Jones algorithm [4], image warping, and ellipse processing—was compared [1]. It claimed that ellipse processing helps an SVM gender classifier achieve a higher performance than using warping of the images or face detection. The benefits of face alignment and normalization were demonstrated [5]. Various over-sampling techniques were also proposed. Training data was enlarged by gathering various web image data [2]. The study presented the importance of the completeness of the training data with respect to race, age, and attire. Augmentation of the training dataset with multiple crops around faces was used to adjust various misalignments [3].

Approaches to gender classification can be mainly categorized into twofold: the non-deep learning techniques and deep learning techniques. Non-linear support vector machines (SVM) had been shown to outperform other traditional non-deep learning methods in gender classification such as linear classifiers, quadratic classifiers, Fisher linear discriminant classifiers, and nearest neighbor classifiers [6]. Adaboost along with modified Local Binary Pattern (LBP) achieved higher performance than rectangular Haar features or wavelet transformations [7]. Gabor filters with Adaboost and SVM classifiers showed superior performance on the Feret database [8]. Textures of small regions in the face were represented by a vector representation of LBP histograms, and then introduced to an SVM classifier [9]. In a similar way, edge histograms for geometric and appearance feature extraction were introduced to a linear SVM [10]. Recently, Multi-Level Local Phase Quantization (ML-LPQ) features, which are texture descriptors, were extracted from normalized face images with non-linear Radial Basis Function (RBF) kernel SVM for gender classification [11]. A combination of elliptical local binary patterns and local phase quantization was also used as a feature extraction technique [12].

Deep learning-based methods have shown significant improvement in performance of gender classification, due to the capability to represent complex variations in facial images. A deep convolutional neural network with five pairs of convolutional layers was proposed [2]. The first four pairs of the convolutional layers are each followed by a max-pooling layer, whereas the last pair is followed by an average pooling layer. Then, the series of the convolutional layers are followed by two fully-connected layers and a softmax output layer. It also suggested that stacks of filters of smaller sizes in the convolutional layer extract more complex nonlinear features from larger receptive fields with less parameters than that of larger filters [2]. A convolutional neural network with three convolutional layers, each followed by a max-pooling layer, and followed by two fully-connected layers and a softmax output was proposed for gender and age classification [3]. Rectified linear units for activation and dropout for regularization were considered in both [2] and [3]. In addition, a local response normalization layer after each of the first two convolutional layers for further generalization was applied [3].

3 Deep Learning

In this section, we briefly introduce neural networks and convolutional neural networks in general. Then, we describe a deep learning architecture that has frequently been used for gender classification.

3.1 Background in Neural Networks

Initial models of deep neural networks had been proposed in the 1960s to the 1970s. However, infeasible computational complexity for training the deep neural network models had made the implementation practically inefficient, despite the development of gradient descent algorithms by backpropagation, which are efficient optimization algorithms for training deep neural networks [13].

During the past several decades, the development of Convolutional Neural Networks (CNN) and GPU acceleration have led to major advances in deep learning models. It allows deep learning to perform far better than state-of-the-art machine learning algorithms such as support vector machines in many computer vision tasks and even better than humans in certain cases of visual pattern recognition tasks [2, 13–15].

Inspired by biological neural systems, a neural network typically includes a number of ordered layers of neurons. The layers include the following three: (1) an input layer, (2) hidden layers, and (3) an output layer. The input layer, the leftmost layer of the model in Fig. 1, introduces data into a neural network. Hidden layers consist of a number of layers, where inputs are transformed to next layers by non-linear transformation. Then, the output layer, the rightmost layer of the network in Fig. 1, outputs the results of the neural network for classification problems.

In neural networks, neurons of hidden layers receive inputs from the neurons of the adjacent previous layer. Suppose that a neural network has L layers and there are N_l neurons in the lth layer. The linear combination of the outputs from the multiple neurons in the previous layer with weights and a bias produces a weighted input for a neuron in the next layer. The weighted input z_j^l of the jth neuron of the lth layer of the neural network is given:

$$z_j^l = \left(\sum_{k=1}^{N_{l-1}} \left(w_{jk}^l \alpha(z_k^{l-1}) \right) \right) + b_j^l, \tag{1}$$

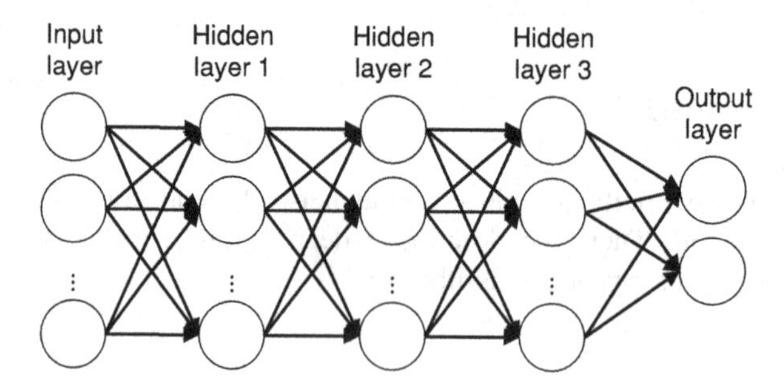

Fig. 1 Overview of neural networks

where w_{jk}^l is the weight connecting from the kth neuron of the $(l-1)$th layer to the jth neuron of the lth layer, and b_j^l is the bias of the jth neuron in the lth layer. In particular, a layer of a neural network, where each neuron of the layer receives outputs from all the neurons of the previous layer or all the inputs from the input layer, is called *a fully-connected layer*. All of the hidden layers and the output layer in Fig. 1 are fully-connected layers.

The outcome of the activation function, $\alpha(z_k^{l-1})$, is referred to as an activation of the kth neuron of the $(l-1)$th layer, which is the final representation of the neuron. Hyperbolic tangent, rectified linear units (ReLU), and sigmoid function are predominant activation functions. Among them, we considered a sigmoid function for the activation function in the gender classification in this study. The sigmoid activation is given by:

$$\alpha(z_j^l) = \frac{1}{1 + e^{-z_j^l}}. \tag{2}$$

A fully-connected layer that uses a logistic or sigmoid function as the activation function of the neurons is called *a sigmoid fully-connected layer*. We follow the convention of using the same activation function on all the neurons of any particular layer.

In supervised learning classification problems, each neuron in the output layer of a neural network typically corresponds to a category of the classification problems. In our neural network model, we used a softmax function for the activation function in the output layer, defined as:

$$\sigma(z_j^L) = \frac{e^{z_j^L}}{\sum_k e^{z_k^L}}, \tag{3}$$

where $\sigma(z_j^L)$ is the softmax activation that estimates the probability of the class corresponding to the jth output neuron, and $e^{z_j^L}$ is the exponential of the weighted input in the jth output neuron. Given the activations of the neurons in the output layer computed by an activation function, the neural network predicts the class j^* corresponding to the output activation with the highest activation as:

$$j^* = \underset{j \in N_L}{\arg\max}\, \sigma(z_j^L). \tag{4}$$

For gender classification, we consider a binary classification problem that predicts male or female from facial images, i.e., there are two neurons in the output layer. We suppose that training data contains pixels of labeled gray-scale facial images, i.e., $\{x | 0 \leq x \leq 255\}$ for our gender classification problem.

In neural networks, there are parameters to be optimized for the optimal model, such as weights and biases on each layer. Back-propagation with gradient descent algorithm is often employed to optimize the parameter using the gradient of the cost function. The cost function involves the output activation of the neurons in the output layer. We used a negative loglikelihood cost function, which is generally used with the softmax output. The negative loglikelihood cost, C, is defined as:

$$C = -\log\left(a_y^L\right), \tag{5}$$

where a_y^L is the softmax output activation in the output neuron corresponding to the class of the ground truth of the training example.

3.2 Convolutional Neural Networks

Convolutional Neural Networks (CNN) introduce additional hidden layers, Convolutional Layers (CL), in neural networks. The convolutional layers identify patterns in small patches of an image by using filters to extract relevant features from the image irrespective of where in the image that pattern occurs. Therefore, convolutional layers can take into account image representations of a set of pixels, while fully-connected layers only consider pixel values. For gender classification, CNN has a strength in identifying facial components such as a nose, eyes, lips, eyebrows, and hair, especially when they do not occur at the same pixel locations in all the training images.

In CNN, more than one convolutional layer is often considered for image classification problems. The first convolutional layer extracts translation-independent features from images. Then, the successive convolutional layers build hierarchical abstractions from these features enabling deep learning.

A convolutional layer consists of several filters of the same filter-size. Each filter is associated with a separate set of neurons, which can be arranged into a sublayer of the convolutional layer. A filter extracts a particular pattern or a feature from an image. Filters move in steps along the image overlaying a rectangular region of the image (typically a square region). The square region that a filter covers while moving along the image is called a *receptive field*, and the receptive field is associated to a neuron in a convolutional layer. The distance in pixels that a filter moves during a single step is called the *stride length*.

The weighted input of a neuron in a convolutional layer is computed from the inputs of its receptive field and the weights of the filter. Suppose that there is a p-by-p filter in the first convolutional layer, where $w_{m,n}$ indicates the weight on the mth row and the nth column of the filter. Then, the activation, $a_{j,k}$, from the neuron in the jth row and the kth column of the hidden sublayer corresponding to the filter is given by:

$$a_{j,k} = \sigma\left(b + \sum_{m=0}^{p-1}\sum_{n=0}^{p-1}\left(w_{m,n} i_{j+m,k+n}\right)\right), \tag{6}$$

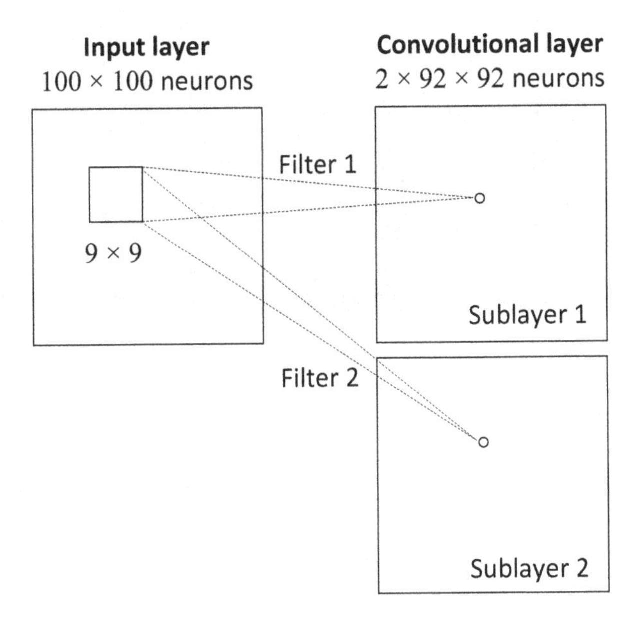

Fig. 2 Filters on convolutional layers

where $\sigma(\cdot)$ is the activation function of the convolutional layer, b is the bias of the neuron, and $i_{j+m,k+n}$ is the pixel input at row $j + m$ and column $k + n$ of the image. Note that the weights and the bias are the same for all the neurons of the sublayer corresponding to the filter. For instance, Fig. 2 shows the convolutional layers with a 100-by-100 image and two 9-by-9 filters. On the left side of the figure, the inside square shows a 9-by-9 receptive field where the two filters are applied. Two sublayers of neurons corresponding to the filters are shown on the right side of the figure, where the circle represents a neuron.

4 Data Sets

We used the iMDB and Wikipedia datasets to train the deep neural networks. The iMDB dataset contains 460,723 facial images (with gender labels) of film stars, predominantly Hollywood actors and actresses, and the Wikipedia dataset includes 62,328 of celebrities from various fields, such as sports, politics, social events, and the film industry.[1] The dataset provides metadata such as a face score, a second face score, age, and gender labels on each image. Images with only one frontal face have high face scores, while those with more than one face or profile faces have low scores. The second face scores indicate how clearly a second face is shown in the image. Figure 3 shows some examples of the datasets.

[1]The original datasets are available at https://data.vision.ee.ethz.ch/cvl/rrothe/imdb-wiki/.

Fig. 3 Examples of the iMDB and the Wikipedia datasets

Table 1 The distribution of male and female images on iMDB and Wikipedia

	Male		Female	
	Number	Percentage (%)	Number	Percentage (%)
iMDB	14,370	43.35	18,777	56.65
Wikipedia	1830	57.03	1379	42.97

We further selected the facial images with a single face, which is mostly frontal. To achieve it, we chose images with a face score equal to or above 4.5 in the iMDB dataset and equal to or above 5 in the Wikipedia dataset, where the second face score indicated no other face. Finally, the iMDB and Wikipedia dataset contain 33,147 and 3209 facial images respectively (see Table 1).

We considered equally distributing the data between male and female. Among the facial images, we randomly chose 14,000 images of each gender from iMDB (total 28,000 images) and 1350 from Wikipedia (total 2700). Then, the original color facial images were converted into 100-by-100 gray-scale images. The final datasets of iMDB and Wikipedia and all of the materials related to our experiments can be downloaded at http://datax.kennesaw.edu/imdb_wiki.

5 Deep Learning Architecture

Deep neural networks can be flexibly constructed in many ways with various numbers of layers and neurons in the layers. Notwithstanding of the increasing business demands and research endeavors in gender classification from facial images, the optimal architecture of deep neural networks for gender classification, however, have seldom been studied. Thus, we provide comparative analysis of the effects with different architectural settings based on a conventional deep neural network. We focus on various numbers of convolutional layers and fully-connected layers for obtaining the optimal gender classification.

We measured the performance of the neural network classifiers with the following performance metrics: the accuracy of prediction, the female and male sensitivity and precision, the informedness, and the average time taken to complete an epoch of training. The performance were computed on the iMDB and Wikipedia image datasets.

5.1 Experimental Settings

Our convolutional neural networks consist of a series of convolutional layers, followed by fully-connected layers, and softmax outputs as most studies have used for gender classification [2, 3]. We considered the following four different CNN architectures in the comparative experiments: (1) two convolutional and two fully-connected layers (C2F2S), (2) two convolutional and four fully-connected layers (C2F4S), (3) three convolutional and two fully-connected layers (C3F2S), and (4) three convolutional and four fully-connected layers (C3F2S). Each convolutional layer is followed by a max-pooling layer. We predefined the number of filters and the filter sizes in the convolutional layers and the numbers of neurons in the fully-connected layers empirically. The details of the four CNN architectures are described in Table 2. The constant learning rate was set to 0.1 and a mini-batch size of 100 training data was considered for stochastic gradient descent in all of our experiments. The stride length was set to one in all of our convolutional layers. We did not consider regularization in our experiments.

Table 2 The architectures of the convolutional neural networks used in the gender classification experiments

	Convolutional layers	Fully-connected layers (numbers of neurons)	Output layer
C2F2S	Two layers with 64 filters of filter-size 9×9, 32 filters of filter-size 3×3 (with MP of pool-size 2×2)	Two layers with 5000 and 500 neurons	Softmax with 2 neurons.
C2F4S	Two layers with 64 filters of filter-size 9×9, 32 filters of filter-size 3×3 (with MP of pool-size 2×2)	Four layers with 5000, 1000, 500, and 100 neurons	Softmax with 2 neurons.
C3F2S	Three layers with 64 filters of filter-size 9×9, 32 filters of filter-size 3×3, 16 filters of filter-size 3×3 (with MP of pool-size 2×2)	Two layers with 5000 and 500 neurons	Softmax with 2 neurons.
C3F4S	Three layers with 64 filters of filter-size 9×9, 32 filters of filter-size 3×3, 16 filters of filter-size 3×3 (with MP of pool-size 2×2)	Four layers with 5000, 1000, 500, and 100 neurons	Softmax with 2 neurons.

MP max-pooling layer

The convolutional neural networks were implemented in python on an IPython notebook using numpy, pandas, matplotlib, OpenCV, and Theano. The gender classification experiments were carried out on an Ubuntu 14.04.5 server using one Tesla M40 GPU core using Cuda, CuDNN and with CNMeM enabled.

5.2 *Experimental Results*

Intensive experiments for the comparative analysis of the four convolutional neural network architectures (C2F2S, C2F4S, C3F2S, and C3F4S) were carried out with a large number of face images from the iMDB and Wikipedia datasets. The iMDB dataset was used for both training and validation, while the Wikipedia dataset was used for only testing.

We compared the performance of the convolutional neural networks by measuring prediction accuracy, sensitivity, precision, and informedness, which are defined by (7)–(12), through the tenfold cross-validation. The iMDB dataset was split into ten groups of equal size, where the samples within the nine groups are used for training and one is for validation. On each fold of the cross validation, the entire Wikipedia dataset was used for testing. We computed the accuracy of the gender classification with the male and female images together (7), whereas sensitivity (8) and precision (9) were measured separately. Note that the numbers of male and female images were equally distributed within the training and test datasets. Informedness represents the probability that the classifier makes an informed guess as compared to random guessing [16].

$$\text{Accuracy} = \frac{\text{\# of facial images correctly classified}}{\text{\# of facial images}} \tag{7}$$

$$\text{F-sensitivity} = \frac{\text{\# of female facial images correctly classified}}{\text{\# of female facial images}} \tag{8}$$

$$\text{F-precision} = \frac{\text{\# of female facial images correctly classified}}{\text{\# of facial images classified as female}} \tag{9}$$

$$\text{M-sensitivity} = \frac{\text{\# of male facial images correctly classified}}{\text{\# of male facial images}} \tag{10}$$

$$\text{M-precision} = \frac{\text{\# of male facial images correctly classified}}{\text{\# of facial images classified as male}} \tag{11}$$

$$\text{Informedness} = (\text{F-sensitivity} + \text{M-sensitivity}) - 1 \tag{12}$$

According to the experimental results in Table 3, C3F2S shows the best performance among the four convolutional neural networks. The informedness and the accuracy of C3F2S were 91.76 ± 0.59 and 95.87 ± 0.30 respectively on the validation data (iMDB) in the tenfold cross-validation. The results were also consistently shown on the test with the Wikipedia dataset as 83.54 ± 0.95 and

Table 3 Experimental results on the validation data and the test data by ten-fold cross-validation

		C2F2S	C2F4S	C3F2S	C3F4S
Validation (iMDB)	Informedness (%)	90.87 ± 0.94	90.36 ± 0.63	**91.76 ± 0.59**	91.11 ± 0.76
	Accuracy (%)	95.42 ± 0.48	95.14 ± 0.33	**95.87 ± 0.30**	95.53 ± 0.39
	F-sensitivity (%)	95.43 ± 0.89	95.46 ± 1.24	**96.03 ± 0.71**	95.56 ± 1.11
	F-precision (%)	95.43 ± 1.01	94.83 ± 1.59	**95.71 ± 0.79**	95.51 ± 1.23
	M-sensitivity (%)	95.44 ± 0.93	94.91 ± 1.42	**95.73 ± 0.72**	95.54 ± 1.14
	M-precision (%)	95.41 ± 0.96	95.45 ± 1.34	**96.03 ± 0.76**	95.54 ± 1.22
Test (Wikipedia)	Informedness (%)	82.97 ± 0.67	81.82 ± 0.93	**83.54 ± 0.95**	82.62 ± 0.20
	Accuracy (%)	91.58 ± 0.40	91.14 ± 0.46	**92.31 ± 0.49**	91.62 ± 0.54
	F-sensitivity (%)	95.00 ± 1.02	94.67 ± 1.63	**95.17 ± 0.66**	94.29 ± 1.69
	F-precision (%)	86.91 ± 1.67	85.85 ± 3.21	87.39 ± 1.62	**87.43 ± 2.34**
	M-sensitivity (%)	87.96 ± 1.23	87.15 ± 2.33	**88.37 ± 1.25**	88.37 ± 1.76
	M-precision (%)	95.41 ± 1.04	95.08 ± 1.71	**95.55 ± 0.69**	94.64 ± 1.83

The bold font represents the highest accuracy on the measurement

92.31 ± 0.49, respectively. The overall performance with the Wikipedia dataset was lower than with the iMDB dataset, since only the iMDB dataset was used for training the convolutional neural networks and the Wikipedia dataset was considered only for testing. Although both iMDB and Wikipedia contain facial images, the performance gap may be caused by the difference of the scopes that the datasets originally cover.

Moreover, we observed the trajectory of the accuracies of the four convolutional neural networks with the validation and test datasets over epochs. Neural network classifiers are iteratively trained over several epochs until the cost function converges. Note that an epoch is a single iteration of training on a neural network classifier with the whole training data input. In the experiments, the trainings were performed with 70 epochs for C2F2S and C2F4S, and 170 epochs for C3F2S and C3F4S.

C2F2S and C3F2S converged to the solution much faster than C2F4S and C3F4S, respectively (see Fig. 4). Both C2F2S and C3F2S quickly converged to the optimal solution at 20 epochs, while C2F4S and C3F4S reached at 50 and 120 epochs respectively. It shows that training fully-connected layers causes more computational complexity than convolutional layers.

We also measured the computational cost of training the convolutional neural networks. In Table 4, it appears that a convolutional neural network with three convolutional layers trains faster than a convolutional neural network with two convolutional layers. The average training times on an epoch for C2F2S and C2F4S are more than 600 s, but 150 s on C3F2S and C3F4S. That is, a single training of the convolutional neural networks with two convolutional layers (with 70 epochs) takes more than 12 h, whereas three convolutional layers (with 170 epochs) needs about 7 h. It is because a convolutional layer reduces the number of parameters to train for the neural network by using shared weights. The consecutive convolutional layers, each followed by a max-pooling layer, then lessen the number of parameters to train even further.

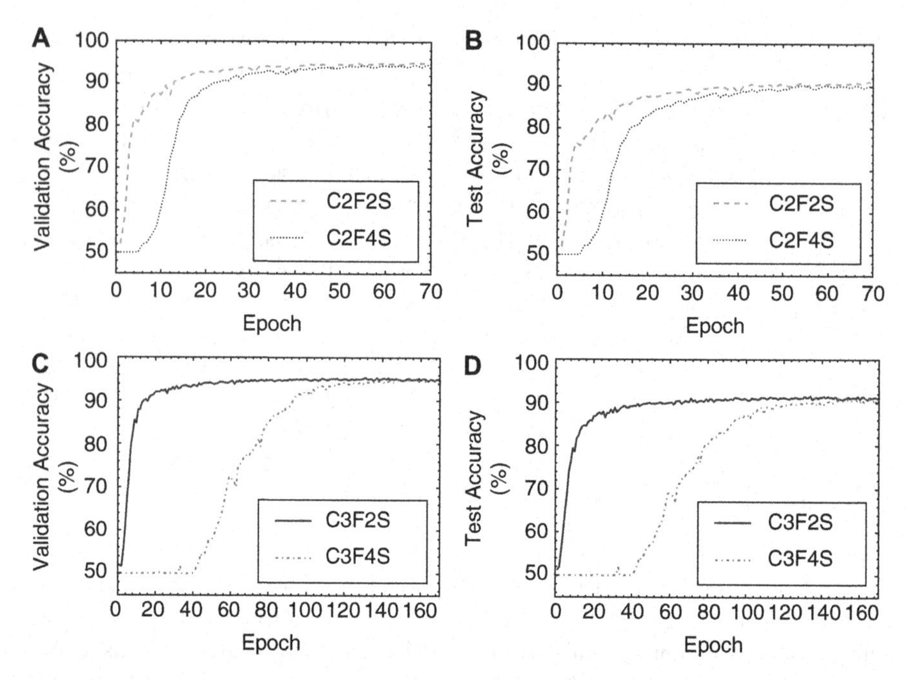

Fig. 4 Accuracy on the validation data and the test data over epoch on (**a**)–(**b**) C2F2S and C2F4S, and (**c**)–(**d**) C3F2S and C3F4S

Table 4 The average training time on an epoch

Architecture	Average time taken to complete an epoch on training (s)
C2F2S	638.89 ± 3.20
C2F4S	667.90 ± 9.33
C3F2S	**150.68 ± 1.09**
C3F4S	179.58 ± 0.12

The bold font represents the lowest training time

Preliminary experiments extending to the fully-connected networks of six and eight layers along with convolutional layers were also conducted. However, higher numbers of fully-connected layers produce no more better performances than two fully-connected layers in gender classification problem.

The averaged images of correctly classified and misclassified facial data with C3F2S are illustrated in Fig. 5. The correctly classified facial images in Fig. 5a and c indicate more clear faces than the misclassified images in Fig. 5b and d. The misclassified images have noise in the background, which may hinder the performance. Interestingly, it is observed that the misclassified female face has short hair, looking similar to the average correctly classified male face.

Examples of misclassified male and female facial images are shown in Fig. 6. Most misclassified male had long hair or dark background that makes it difficult

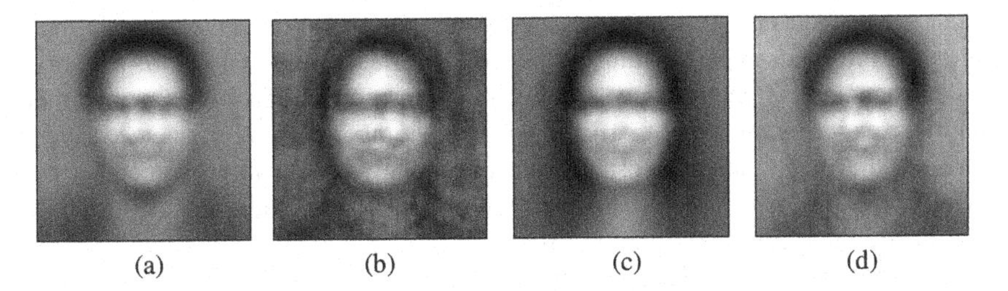

Fig. 5 Averaged facial images of (**a**) correctly classified male, (**b**) misclassified male, (**c**) correctly classified female, and (**d**) misclassified female with C3F2S

Fig. 6 Examples of (**a**) misclassified male and (**b**) female facial images

to discern hair from the background, as shown in Fig. 6a. On the other hand, the misclassified female images show short hair as shown in Fig. 6b, which may predominantly cause the misclassified results.

6 Conclusion

Automatic gender classification system is essential in a wide range of real-world applications and research. We present a deep learning approach to gender classification from facial images in this chapter. A detailed description of neural networks and convolutional neural networks for gender classification are provided. Moreover, the comparative analysis was conducted by intensive experiments to help one to choose the optimal architecture of a convolutional neural network for gender classification.

Our experiments may indicate that increasing the number of convolutional layers decreases the time taken to train the network in a convolutional neural network, while enhancing the performance of a deep learning gender classifier. On the other hand, our experiments show that increasing the number of fully-connected layers increases the computational cost to train the model without improvement in the performance. This is in agreement with the findings from other studies in gender classification using deep learning [2, 3].

Even the simple convolutional neural network used in our experiment, which consists of three convolutional layers and two fully-connected layers, produces a good gender classification performance of 95% accuracy.

References

1. Lu, H., Lin, H.: Gender recognition using Adaboosted feature. In: Third International Conference on Natural Computation (ICNC 2007) (2007)
2. Deng, Q., Xu, Y., Wang, J., Sun, K.: Deep learning for gender recognition. In: 2015 International Conference on Computers, Communications, and Systems (ICCCS), pp. 206–209 (2015) [Online]. Available http://ieeexplore.ieee.org/document/7562902/
3. Levi, G., Hassner, T.: Age and gender classification using convolutional neural networks. In: 2015 IEEE Conference on Computer Vision and Pattern Recognition Workshops (CVPRW), Boston, MA, pp. 34–42 (2015)
4. Viola, P., Jones, M.: Rapid object detection using a boosted cascade of simple features. In: Proceedings of the 2001 IEEE Computer Society Conference on Computer Vision and Pattern Recognition (CVPR) 2001, vol. 1, pp. I-511-I-518 (2001). https://doi.org/10.1109/CVPRW.2015.7301352
5. Ng, C., Tay, Y., Goi, B.M.: Recognizing human gender in computer vision: a survey. In: PRICAI 2012: Trends in Artificial Intelligence. Lecture Notes in Computer Science, vol. 7458, pp. 335–346 (2012)
6. Moghaddam, B., Yang, M.H.: Learning gender with support faces. IEEE Trans. Patt. Anal. Mach. Intell. **24**(5), 707–711 (2002)
7. Verschae, R., Solar, J., Correa, M.: Gender classification of faces using Adaboost. In: Progress in Pattern Recognition, Image Analysis and Applications, pp. 68–78 (2006)
8. Lin, H., Lu, H., Zhang, L.: A new automatic recognition system of gender, age and ethnicity. In: 2006 6th World Congress on Intelligent Control and Automation, vol. 3, pp. 9988–9991 (2006) [Online]. Available http://ieeexplore.ieee.org/lpdocs/epic03/wrapper.htm?arnumber=1713951
9. Lian, H.-C., Lu, B.-L.: Multi-view gender classification using multi-resolution local binary patterns and support vector machines. Int. J. Neural Syst. **17**, 479–87 (2007) [Online]. Available http://www.ncbi.nlm.nih.gov/pubmed/18186597
10. Ardakany, A., Jou la, A.: Gender recognition based on edge histogram. Int. J. Comput. Theor. Eng. **4**, 127–130 (2012)
11. Bekhouche, S.E., Ouafi, A., Benlamoudi, A., Taleb-Ahmed, A., Hadid, A.: Facial age estimation and gender classification using multi level local phase quantization. In: 3rd International Conference on Control, Engineering and Information Technology, CEIT 2015, pp. 3–6 (2015)
12. Nguyen, H.T.: Combining local features for gender classification. In: Proceedings of 2015 2nd National Foundation for Science and Technology Development Conference on Information and Computer Science, NICS 2015, pp. 130–134 (2015)
13. Schmidhuber, J.: Deep learning in neural networks: an overview. Neural Netw. **61**, 85–117 (2015) [Online]. Available http://dx.doi.org/10.1016/j.neunet.2014.09.003

14. Sun, W., Su, F.: Regularization of deep neural networks using a novel companion objective function. In: International Conference on Image Processing (ICIP), pp. 2865–2869 (2015)
15. Nielsen, M.: Neural networks and deep learning (2015) [Online]. Available http://neuralnetworksanddeeplearning.com/index.html
16. Powers, D.M.W.: Evaluation: from precision, recall and F-factor to ROC, informedness, markedness & correlation. Technical Report SIE-07–001. School of Informatics and Engineering Flinders University, Adelaide (2007)

Social and Organizational Culture in Korea and Women's Career Development

Choonhee Yang and Yongman Kwon

Abstract The biggest challenge faced by the Korean labor market today is its labor shortage due to the rapid decrease in its productive population. Korea's productive population dwindles from 2017, reaching its peak on 2016. The speed of the productive population is decreasing faster than any other country. Korea's compressed economic growth during its industrialization process was made possible by an abundance of high quality labor but since then a low birthrate and its rapidly aging society have caused a continued decrease that ultimately led to its current shortage.

1 Introduction

The biggest challenge faced by the Korean labor market today is its labor shortage due to the rapid decrease in its productive population. Korea's productive population dwindles from 2017, reaching its peak on 2016. The speed of the productive population is decreasing faster than any other country. Korea's compressed economic growth during its industrialization process was made possible by an abundance of high quality labor but since then a low birthrate and its rapidly aging society have caused a continued decrease that ultimately led to its current shortage [1].

Sustainable economic growth requires a stable source of labor and the answer to Korea's problem could be found in its population of highly-educated women that remains largely unused. The question then should be asked, what is the cause behind under developed such a rich source of human resources?

The labor shortage is an issue that must be resolved in one way or another. Korea's male labor force which has been developed and utilized as the main source of labor is no longer capable of powering economic growth. Thus, the increase of highly-educated female labor force entering the labor market and the continued utilization of their labor is necessary for Korea to increase its productive population. Female employment is an area where Korea lags behind not just in comparison to other aspects for women but also when compared to other OECD countries.

C. Yang (✉) • Y. Kwon
Department of Business Administration, Namseoul University, Cheonan, Choong-Nam, South Korea
e-mail: chyang@nsu.ac.kr

S.C. Suh, T. Anthony (eds.), *Big Data and Visual Analytics*,
https://doi.org/10.1007/978-3-319-63917-8_4

Korea's society culture can be highly correlated with women's career development, such as patriarchal male-dominated culture, high-context society, military culture, and the socialization of gender role etc. The patriarchal, male-dominated culture of the Korean society has created sexist stereotypes against women that are strengthened and spread within the society via language. Its social norms have caused the socialization of discriminatory gender roles, the root of the prevalence of sexual discrimination against highly-educated women in employment and human resources management, such as career development, promotion, and other workplace opportunities as various statistics show.

Also, the political situation of Korea, as divided in to South and North Korea due to the Korean-war from 1950 to 1953, has made soldiers rule the nation for a long period. Military culture influences personnel customs from all over the society and also in the organization of a firm.

While Korea has one of the highest rates of female college enrollment in the world, its level of female economic activity is one of the lowest among OECD countries. The reality is that the percentage of female management staff and female CEOs at the top 30 firms in Korea remain very low despite the fact that there are many highly-educated women due to the difficulty in entering the labor market and the career interruption caused by pregnancy, child birth, and child rearing.

This study attempts to look at the characteristics of Korea's social and organizational culture and the social stereotypes pertaining to women and analyze the status quo and quality of female employment as seen via statistics. Statistics by the national statistics office such as college enrollment rates, economic activity rates, career-interrupted women ratio, non-regular employee ratio, the percentage of female management staff and its ratio to all workers by gender have been used as the main source material to understand where female labor stands in Korea today.

2 Korean Social Culture

Korea's abundant human resources have made a country with no natural resources to the position of the ten largest economies in the world. Through this process, some aspects have led positive economic developments, such as rapid economic development, while some other aspects have deduced unfavorable results. There are countless examples of these social phenomena; sturdiness of patriarchy, short-term achievement principle, collectivism that oppress freedom of an individual, military culture of uniformity and commands, and the culture which puts emphasis on context that law and order do not operate.

2.1 Korea's Patriarchal, Male-Dominated Culture

A Country's traditional culture is the primary and most influential factor in developing gender role biases. Women's status was relatively high during the Three Kingdom Era and Koryo but the adoption of Confucian values during the Chosun Period placed women below the starts of men for 500 years during the Chosun Dynasty [2]. Confucianism was upheld in the Korean society for half a millennium during the Chosun Dynasty and as such, the Confucian traditions still remain deep-rooted in modern day Korea. The tradition emphasizes values such as filial piety and respect for the elderly but at the core of it lies a patriarchal culture, where the father is the head of all family members whose opinions cannot be disputed by the family.

This patriarchy can be detailed through the process of women's formation of labor forces, after the Korean War of the 1950s, during the times of modernization (1960s and 1970s). According to a research held by Mikyung Jang, Korean women's labor in the 1960s and 1970s, early years of modernization (industrialization) in Korea, is shown as women from a society of agriculture becoming a 'housemaid' at homes in the cities. Married women tend to adhere to farming. With industrialization advancing, female workers in the countryside had to work for a family living in the city and take responsibility for the upkeep of the family.

Furthermore, women had to give up their higher education, mainly because of the social awareness that women need not education and that getting married once they support their new families is their responsibility. At that time, though women wanted to develop oneself through education, those women had to sacrifice themselves for the sake of their brothers' education, in times of poverty [3]. Regarding that men mainly need education and investment toward women's education is needless had been prevailing. Women's sacrifice for male reproductive systems resulted in 'Boys' preference' and consequently made patriarchal culture put down its roots in times of industrialization and capitalism. Not to mention the situation on the eldest daughter. Eldest daughters, considered as 'Son of the Family, felt high responsibility as the head of the household [3, 4]. The Confucian family-ism ideology had strengthened women's employment on city factory [5]. During this time, women weren't so unsatisfied of themselves not getting education.

Time passed by and Korea has achieved a rapid economic development, while still in the late twentieth century, the late period of the industrialization age, had women been unable to cultivate a sense of career women, who are self-contained of their lives, due to the dominant men-provider model. Meanwhile, women who once worked at farms but worked in factories had to endure their poor, discrepant environment while being depreciated by the term 'Gongsooni'(a degrading expression on the female factory worker) and sending almost half of their wage to their families. The driving force of these women was that they could bring an end to their work once they get married. Eventually, the dichotomous concept that men work for the society while women work for private aspects such as family and households was prevail. This worked, in later times, as a hindrance for women to achieve higher status [6]. Korean society culture in this time had a tendency of viewing women who still works after marriage matters as abnormal, or even as pity

[3]. Eventually, the concept that men are productive individuals doing labor and women are consumers after marriage was settled.

In the early ages of industrialization, women got their jobs using their human network regarding their hometown and societal relationships. For easy employment by association and to be placed at relatively plain jobs, women bribed personnel managers, worked as a housemaid for some period with no pay, or they were hired as unskilled workers. Some qualification for women employment in the 1960s were mostly, age, marital status, academic ability, and external appearances. Sometimes, young women of age lower than 14 were employed. This is also shown in economic activity rates in the 1960s and 1970s. Economic activity rates of age 15–24 in the 1960s reached 40 50%, and in 1975, economic activity of 14 years old accounted for 17.7%. In the 1970s, unmarried women accounted for 77.5% out of total women workers.

Such male-dominated patriarchal traditions at home were also extended to other organizations, such as the workplace. Thus, there is an evident gender gap in terms of the quality of employment as well as the human resource management system in the workplace. Male and female gender roles, the social role played by each gender according to the actions and attitudes deemed appropriate by the society's culture, are learned via a socialization process that occurs within one's social groups, such as the family, school, community, and workplace. Socialization is the process by which one acquires the actions, beliefs, standards, motivation, and so on that one's affiliated culture upholds as valuable [7].

The socialization of gender roles begins in the family with the parents. The family is a microcosm of the society in that it contains the same political and economic structures, and it is here that the family is the first male-dominated social organization to which one is exposed. And institutional education is the influence to shape biases and human resource practices after one enters the workforce is the determining element. Unequal development of women leaders is caused by the educational environment requiring role division according to gender. Korean Women's Development Institute (KWDI) inspected the sixth curriculum edition of school textbooks (102 books at the elementary level, 26 books at the middle school level, and 23 books at the high school levels for a total of 151 books) and found that the textbooks contained traditional elements of gender discrimination [8]. Contrasting experiences were noted by a lecturer at a coed university who originally used to work at a women's university. Male students, unlike female students, actively sought leading roles within their team of peers. Male team leaders were most vocal and showed the tendency of dismissing their peers' opinions, especially if they were from female students. Meanwhile, female students tended to be less confident, spoke with a softer voice, and less active in sharing their thoughts. Instances of sexist conclusions being drawn were also observed [9]. From such observations, it can be concluded that students learn the social stereotypes and gender roles of men leading and women following when male and female students discuss to reach a decision. Once learned, the social norms regarding gender roles will remain deep-rooted in oneself, which is why even highly-educated women will still be under the influence of the patriarchal culture in their family.

2.2 Social Norms and the Socialization of Gender Roles

The acquisition of language is how one starts using communication as a learning tool, which points to verbalism as one tangible proof of the social conventions on women. The language used reflects the internalized values and attitudes held by the speakers, and the Korean language is marked with openly derogatory verbalism against women. In a survey of Korean and American university students and their exposure to sexist language, Korean students replied with a higher frequency and variety of discriminatory language experienced than American students. The majority of the terms Korean students had been exposed to were simple phrases full of sexism such as "how dare a woman," and "that's what women do" [10].

Language spreads the social norms of male-domination within society including the workplace and the women themselves, and the acceptance of such discrimination as natural and reasonable can lead to a vicious cycle against women in terms of work ethic and success in the workplace. According to the socialization theory of gender roles, one's sex in relation to the society is central to the formation of one's gender identity. Thus, the discriminatory socio-cultural structure deep-rooted in the society causes the low status of women in the workplace and the low quality of female employment even in the knowledge- information society. Korea has developed the lower status of women in the workplace and the lower quality of female employment.

According to the World Values Research Institute conducted among OECD countries on the gender gap in employment during severe unemployment crises, Korea had the biggest gap at 34.5% in the belief that male applicants should be prioritized. Following Korea was Poland was at 31.1%, Chile at 27.8%, Japan at 27.1%, and Mexico at 25.1%. In contrast, countries with a high level of female social participation were found on the lower end, with Sweden at the lowest at 2.1%, Norway at 6.5%, and the US at 6.8% [11].

Furthermore, the sexist attitude and the gender gap in employment were found to have a correlation, with a 0.437 ($r = 0.437$) the correlation coefficient value between the two variables. Korea had the largest gender gap at 22.4%, followed by Chile and Mexico. This signifies that the bigger the gender gap in employment, the higher the numerical value of the country's sexist attitude.

3 Korea's Economic Indicator and Current State of Women Labor

Of all the economic indicators of Korea, it is shocking that Korea's productive population decreases for the first time from 2017. This is because of Korea's low birthrate and Korea is becoming an extreme aging society. To supplement or buffer the dramatic drop of its productive population, we have to highly educated women, which is in abundance. The recent rate of enrollment to higher education institutions of women is same or higher than that of men. However, women's activity rate is far below and career discontinuity of married women is much higher than other countries in OECD. The solution to this problem is to put off retirement and effectively work out with women labor.

3.1 Korea's Productive Population and Its Competence

There is the prediction that Korea's population structure will be following Japan's, and there is also pessimistic opinion that Korea will turn into an aging society faster than Japan in the near future. Korea's population structure follows the orbit of Japan's, by 10 or 15 years of time difference. Japan has already been examining solutions to defend the decrease of its population, and currently in 2017, a significant amount of national policies are in process. According to Statistics Korea's population estimation, we can see the first decrease of Korea's productive population (15–64 years old), and this rate becoming faster starting from year 2017. Meanwhile, the elderly population (age 65 or over) increases rapidly, so by the year 2032, the elderly population will account for more than 26% out of the total population. Korea's productive population of 2015 is 37,440,000, and in 2016, its population will be 37,630,000 hitting the peak and by 2017 it will face a steep decline [1].

Some argue that we can buffer the impact by merely following Japan's strategy. However, Korea's economic phenomenon is distinguished from that of Japan's. Before Korea stepped into an aging society 10–15 years ago, when Japan's productive population has dwindled, surrounding countries such as China and Korea had lots of young consumers to buy Japanese products. However, current point when Korea is aging, other countries are also aging, which brings lack of overseas demand. In long terms, Korea's economic structure will worsen than Japan and China. The decline of its productive population means that Korea's competence will drop, eventually making Korea poor if solutions do not operate.

3.2 Korean Women's Qualitative Employment Indicators

Employment indicators of early industrialization period (1960s–1970s) to look at Korean women's qualitative employment indicators are like the following tale.

In early industrialization period (1960–1970s), women mostly worked for primary industries. By the later period of industrialization (later part of the 1990s), has the rate of third primary industry workers, working in service industries, finance, retailing and insurance firms increased (See Table 1) [12].

The depreciation of women's work occurs inside social relationships rather than in the work itself [13]. Division of labor between the sexes happens from this

Table 1 1960'–1970' Korean women's employment by industry

	(Unit: %)		
Year	Primary industry	Secondary industry	Tertiary industry
1960	69.6	6.4	22.7
1970	59.7	14.7	25.5
1980	46.5	21.9	31.6
1990	20.4	28.0	51.6
1999	13.3	17.4	69.3

moment. Division of labor is not a biological sex but a system in which labor is divided by socio-culturally divided social gender. From early industrialization ages where capitalism has proceed, Korea's unique patriarchy has made a new governance system based on hierarchy of gender and age. Taking a look at the manufacturing industry, men engage in auto industry while women engage in textile industry. This is an example of division of labor by types of business.

Korean women's employment regarding age shows that in 1960s–1970s women's activity rate of age 15–24 is 40 50%. In 1975, activity rate of 14 years old age is 17.7%. In 1970s 89.7% of the factory's manufacturing labor was women. By the size of the enterprise, young women show tendency of getting employed in small and medium sized firms, and women of age 17–19 who didn't break the Labor Standards Law were employed in big sized firms. Because of the tough conditions in small and medium sized firms, some women used their older sister's identification card to get into big sized firms.

By the level of education in 1972, women who graduated in elementary school accounted for 33.6%, middle school graduates for 27.3%, and high school graduates for 10.5%. To look at women by standards of the sizes of the firms and their education, 63.8% of the elementary school graduates were hired in small firms, while 32.3% of the high school graduates were hired in big firms. Korean parents' high enthusiasm for their children education has led women's enrollment to higher education institutions after the 1980s.

Table 2 shows that the college entrance rate for men was 4% higher than that of women in the 1980s, with 34% of women entering university by the mid-1980s. But gender gap in advancement rate continuously decreased and the female entrance rate peaked at 83.5% in 2008 before overtaking its counterpart in 2009 by 0.8% with 82.4% of women and 81.6% of men entering college. As of 2014 the college entrance rate for female students is 74.6%, which is 7% higher than the 67.6% of male students. In terms of graduate school, the female entrance rate is 6.0% as of 2014, which is lower by 1.6% than the male entrance rate at 7.6% [14].

Even though highly educated women take larger percentages, women participate in economic activities less than male in Korea's labor market (Table 3) (Korea National Statistical Office, 2016). Women's activity rate is low, even comparing to other OECD countries (Table 3) [1] and (Table 4) [15, 16].

Table 2 Advance rate of graduates to higher school level

	High school → College and university		College and university → Graduate schools and etc.	
Year	Female (%)	Male (%)	Female (%)	Male (%)
1985	34.1	38.3	6.7	12.1
1990	32.4	33.9	6.4	8.8
1995	49.8	52.8	6.2	9.2
2000	65.4	70.4	6.9	11.1
2005	80.8	83.3	7.5	9.0
2007	82.2	83.3	5.6	7.6
2008	83.5	84.0	5.5	7.3
2009	82.4	81.6	5.9	7.7
2010	80.5	77.6	5.5	7.1
2012	74.3	68.7	5.8	7.1
2013	74.5	67.4	6.0	7.6
2014	74.6	67.6	–	–

Table 3 Economic activity rate in Korea

	(Unit: 1000 persons, %)			
	Economically active population		Labor force participation rate	
Year	Female	Male	Female	Male
1985	5975	9617	41.9	72.3
1990	7509	11,030	47.0	74.0
1995	8397	12,456	48.4	76.4
2000	9101	13,034	48.8	74.4
2005	9860	13,863	50.1	74.6
2010	10,256	14,482	49.4	73.0
2012	10,609	15,071	49.9	73.3
2013	10,802	15,071	50.2	73.2
2014	11,149	15,387	51.3	74.0

Table 4 OECD Economically Participation Rate

	Unit: %		
Year	2013	2012	2011
Korea	55.6	55.2	54.9
Japan	65.0	63.4	63.2
USA	67.2	67.6	67.8
OECD	62.6	62.3	61.8

3.3 Korean Women's Employment Stabilization Indicator

Highly-educated women provide Korea with a rich source of labor. However, women are at a disadvantage in entering the workplace, and those who succeed face career interruption due to pregnancy, child birth, and child rearing. Career-interrupted women refers to married women of age 15–64, who quit jobs due to marriage, pregnancy/child delivery.

As Table 5 illustrates, 20.7% of all married women have discontinued their career as of 2014, and 36.7% of those women are between the 30–39 age bracket. Statistics Korea reported that there are 1,906,000 career-interrupted women in 2016. Of course, this number has reduced by 147,000 (7.1%) than the year before. However, women in their 30s who experienced career discontinuity accounts for 35.6%. This period is when women can make achievement and mature one's career, but have problem to do so because of child rearing and taking care of their family. There are difficulties for them to continue their work, and though assuming there is possibility to gain jobs again, the discontinuity in their 30s is fatal in that they miss their sense of professional expertise [14].

Marriage was found to be the biggest cause of career discontinuity at 45.9% followed by child rearing at 29.2%. (See Table 6) The career discontinuity of Korean women follows an L-shaped curve as many female workers will leave the labor market upon marriage, focus on child rearing for years, until they re-enter the market in their mid-to-late 30s or give up re-entry for good [8].

There are some meaningful changes that we should pay attention to in the statistics of April 2016 to compare with 2013. The percentage of women who faced career discontinuity due to marriage had exceedingly diminished (45.9% → 34.6%). Yet, women who quit their jobs due to pregnancy and child bearing have rather increased (21.2% → 26.3%). The main reasons for married women quitting their jobs comes down to three factors; marriage, pregnancy/child bearing/child rearing. The three factors mentioned accounts for more than 90%. Career discontinuity due to 'taking care of family' accounts for 4.8%, which is similar which child rearing (4.1%). In Korea, family plays a role in which married women give up their jobs.

Table 5 Career-interrupted women in Korea

Year	(Unit: 1000 persons, %) Married women (A)	Non-employed women (B)	Career-interrupted women (C)	C/A (%)
2011	9866	4081	1900	19.3%
2012	9747	4049	1978	20.3%
2013	9713	4063	1955	20.1%
2014	9561	3894	1977	20.7%

Table 6 The reason of career-interrupted women

Year	The reason of career-interrupted Marriage	Pregnancy and childbirth	Infant care	Education of children	
2013	45.9%	21.2%	29.2%	3.7%	
2016	34.6%	26.3%	30.1%	Total 8.9	
				Family care	Education of children
				4.8%	4.1%

The decrease of the number of career broke women can be evaluated positively comparing past statistics, but there is a different reason. Outlook on marriage has changed in general, and we can find the grounds in Korea's long term recession. The general concept that it is good for women to quit their jobs after marriage has disappeared recently. Also, the number of two-paycheck couples has increased due to loans from early times of marriage, which is because of Korea's low-pressure economy and recession. Eventually, economical depression is bringing down the number of career broke women.

Additionally, Korea's peculiar social culture and organizational culture impedes on activation of such policies for unburdening pressure of children education. The Korean government has been launching two main policies to help career-interrupted women back to workplaces, from 2013. Intensifying 'parental leave system' for coexistence of work and family and enlarging 'part-time employment' are those. But, these systems are restricted to public officials, some big-sized firm employees, and for women who works for the first labor market. There are some side-effects in that firms hire fewer women because of the parental leave system enforced by law. Employers of a private firm influenced by economic sense try to earn high revenues in short terms while the government merely recommends parental leave systems without offering any motive for the firms, which is far from business logic. This explains the criticism towards the government that their policies are just for 'exhibition', which does not come out with any feasible solutions. Besides, part-time jobs are similar to contract workers, which usually does not maintain for a long time. Virtually, firms report statistics to the government just for exhibition purposes, hiring one-time jobs and not sustaining them. This is the sociocultural reality for the government to make policies, and it is the reality for career-interrupted women to get back jobs only to face career discontinuity unstable job security.

Forty-one percent of all female workers were found to be on temporary contracts compared to the 26.8% of males, and it was found that 53.5% of all temporary workers were female. There is also difference between the genders in 2013 and this gap did not face change even in the year 2016. Furthermore, a gender gap in income was also found in terms of wage, with female workers only earning 68% of their counterparts in 2013.

4 Female Management Staff and Female CEOs in Korea

It is difficult for women to comeback successfully after career discontinuity due to marriage, pregnancy and child bearing and child rearing. Though there are regulations such as the gender equality law to help career women overcome hardships and become promoted to positions higher than intermediary manager, women rarely succeed to become chief executives in real life.

4.1 *Female Management Staff and Female CEOs in Korea*

A survey of organizations with more than 500 full-time workers and public organizations was carried out that found the female employment rate in the public sector to be 33.61% and 38.48% in the private sector. The ratio of female management staff averaged 11.5% in the public sector and 17.9% in the private corporations, showing that the ratio of female management staff to total workers is comparatively quite small (Table 7).

Table 8 shows that compared to the OECD average of 28.3%, Korea at 9% has one of the lowest percentages of female management staff [15].

In the private sector most female workers were found at or below the middle management level. In corporations with more than 500 full-time workers, 6.6% of the executives were female in 2009. Table 8 compares the percentage of female executives in OECD countries. Norway has the highest percentage of promotions of female workers to the executive level at 36.1%, Sweden in second at 27%, Fin-land at 26.8%, and France in fourth at 18.3% [17].

Based on the 2015 CEO Score, Table 8 presents the ratio of male to female executives in relation to the total number of workers in the 184 listed companies in the top 30 conglomerates in Korea [18].

4.2 *Positioning Gender Equality and Executive Ratio*

Simply noting that the percentage of female executives has increased with time is insufficient in analyzing the present day conditions in the top 30 conglomerates in Korea. What is of importance is determining whether a positioning gender balance has been achieved. The positioning gender balance is achieved when there is an adequately proportionate ratio of executives to the total number workers. However, it can be seen that the ratio for female executives to total workers is 0.08%, while the male ratio is 1.15%—meaning there is a positioning gender imbalance (See Table 9) [18].

Table 7 Percentage of female employment by institution in Korea

	(Unit: %)					
	Women employment ratio			The ratio of female manager		
Year	Total	Public sector	Private firm	Total	Public sector	Private firm
2013	36.04	33.61	38.46	17.02	11.55	17.96
2014	35.2	32.35	36.74	16.62	11.01	17.59
2015	34.67	31.19	35.56	16.09	10.53	17.13

Table 8 Percentage of female managers in major OECD countries

(Unit: %)					
Norway	Sweden	Germany	UK	France	Korea
41	27	13	12	8	6.6

Table 9 Female CEO ratio to the number of employees in Korea

Total employees			Total board member (B)	Sex ratio CEO (C)	Staff officers contrast ratio (B/A)
Sex	Number (A)	Percent			
Male	760,000	75.8%	7618	97.9%	1.15%
Female	166,000	24.2%	163	2.1%	0.08%
Total	926,000	100%	7781	100%	

Table 10 Percentage of female executives by job position in Korea

Marketing job family	Engineer job	Planning job	R&D job	Human resources job
27.1%	19.2%	18.1%	12.4%	5.1%

I sort of classified the proportion of female executives by job division. Female CEOs mainly doesn't belong to core job family such as HR, R7D, and planning job (See Table 10).

5 Conclusion and Future Directions

The abundance of cheap and high quality labor that drove Korea's economic growth no longer exists. Compared to 34 OECD countries, Korea is below the average in its economically active population while at the top in its rate at which the society is aging. Its labor shortage is becoming increasingly exacerbated and its economic growth is quickly losing momentum.

It was found that Korea's predominantly Confucian and male-dominated culture has led to the existence of sexual discrimination against women in the labor market and the socialization of discriminatory gender roles. The thick 'glass-ceiling' it has resulted hinders female workers to further develop their competency and get promoted to management staff and executive positions.

Korea has a high level of college entrance rate for both male and female students. There are more female students have been entering college than males since 2009 and the entrance rate for graduate school shows no great disparity between the sexes. However, its rate of female economic activity and employment is one of the lowest among the OECD countries, meaning that highly-educated women in Korea are unable to utilize their competence in the labor market. And even if they are employed, they are subject to a career interruption due to pregnancy, child birth, and child rearing.

Most cases of discontinuity occurred when the highly-educated women were in the 30–39 age bracket. The most prevalent reason for the interruption was marriage at 45.9% followed by child rearing issues. Female workers were experiencing an L-shaped career interruption curve characterized by a period of discontinuity that ultimately led to them giving up re-entry into the labor market. It was found that there were more temporary female workers than there were male, and there was a gender gap in wages with female workers earning 68% of the male income.

And even if women succeed in overcoming such obstacles, there exists a thick glass-ceiling for promotions to high level positions in Korea. For positioning gender equality the proportionate ratio of female workers to female management staff is more important than that of female executives. Statistics show a significant gender gap in the ratio of female workers to female executives and that of their male counterparts. The lack of support for child rearing lies at the root of Korea's low birthrate and thus policies that reduce the burden of child rearing such as financial support and a quota system for female executives are necessary if Korea's population of highly-educated women are to realize their latent competency in the labor market.

There was also a significantly low percentage of female management staff as well as female executives in the 184 listed companies of Korea's top 30 conglomerates. Also, the ratio of total female workers to female executives was much lower than that of men, which pointed to a positioning gender imbalance.

Economic growth requires labor, and Korea can increase its productive population through its underutilized female labor force. If Korea turns to female labor as one of its solutions, a great change in the society's perception of gender norms will need to occur for the female labor force to grow and develop. A detailed action plan reforming the past traditions of patriarchal and male-dominated organizational culture is required, as well as social support measures for child rearing and children's education if female workers are to remain in the labor market and develop their competence.

Policies such as the quota system for female executives found in developed countries are necessary to enable highly-educated women in their 30s remain in the workplace in face of needs such as child rearing. For example, Norway was able to sufficient increase its productive population by means of its women's employment quota that sees female executives at 30%. Its neighbors in Europe such as Sweden, Germany, and France also have similar systems in place. Korea's aging society is rapidly becoming an aged society, signaling the further decrease of its production population. As statistics show, highly-educated female human resources are the latent growth engine in Korea's economy.

In this study it was proven that the sexually discriminatory stereotypes and the Korean society's organizational culture are stunting the development of the female labor force, and that the development of highly-educated female labor needs to take place before Korea can solve its labor shortage problem. Based on such facts, this study suggests that the Korean society acknowledge the importance of fostering female leadership as a goal and take the necessary action in support of it, using the policies of developed countries as case studies for reference.

Specific education/training manual is needed for future women resources development. First, there should be solutions for gender equality in education throughout kindergarten, elementary school, and middle/high school. Policies should be made to give motives to firms which put great effort on women resources development.

References

1. Korea National Statistical Office: (2016)
2. Yang, C.: Managerial leadership differences by gender. J. Glob. Bus. Netw. **1**(1), 109–115 (2006)
3. Jang, M.: Modernization and women labors in the 1960' and 1970'. In: Economy and Society, vol. 61, pp. 106–134. The Association of Korean Researchers on Industrial Society, Seoul (2004)
4. Kim, B.: The Actual Condition and Analysis of Korean Labor Women-the Sound of Workplace. Jinyangsa, Seoul (1984)
5. Kim, H.: Korea's Modernity and Women's Labor Rights. East Asian Modernity and Gender Politics, Pooreunsasang Publisher, Korean Women's Studies Association Edit. (2002)
6. Kang, Y.: The Changes of Labor Market and Work-Family Relation after 1960' Industrialization. Korea Women's Studies Institute, Seoul (2007)
7. Anderson, M.L.: Thinking About Women; Sociological and Feminist Perspectives. Macmillan, New York (1983)
8. Korean Women's Development Institute (KWDI): Report 2013 (2013)
9. Nah, Y.: The voices of female students in higher education: insights from a women-only class at a coed college. Asian J. Women Stud. **12**(2), 8–10 (2006)
10. Yang, C., Kwon, Y.: Gender role socialization and its determinants of gender discrimination language. Namseoul Univ. Bus. Res. **5**(2), 10–15 (2008)
11. World Values Research Institute: In case of serious employment difficulties, the attitude about giving priority to gender employment among OECD: Leading countries (2009)
12. Korea Labor-Employer Development Foundation: Employee data report, 1966 (2013)
13. Kemp, A.A. (ed.): Women's Work-Degraded and Devalued. Prentice Hall, Englewook Cliffs (1994)
14. Kosis: Statistics Korea, 2014 (2016)
15. OECD: Human development report (2009)
16. OECD: Employment outlook (2014)
17. Meerae Forum 30% Club, Korea gender diversity report (2015)
18. CEO Score: 30 Conglomerates status of women executives, internal data (2015)

Big Data Framework for Agile Business (BDFAB) As a Basis for Developing Holistic Strategies in Big Data Adoption

Bhuvan Unhelkar

Abstract The Big Data Framework for Agile Business (BDFAB) is the result of exploration of the value of Big data technologies and analytics to business. BDFAB is based on literature review, modeling, experimentation and practical application. BDFAB incorporates multiple disciplines of Information Technology, Business Innovation, Sociology and Psychology (people and behavior, Social-Mobile media), Finance (ROI), Processes (Agile), User Experience, Analytics (descriptive, predictive and prescriptive) and Staff Up-skilling (HR). This paper presents the key elements of the framework comprising agile values, roles, building blocks, artifacts, conditions, agile practices and a compendium (repository). The building blocks themselves are made up of five modules: business decisions, Data—technology and analytics, user experience-operational excellence, quality dimensions and people—capabilities. As such, BDFAB exhibits an interdisciplinary approach to Big Data adoption in practice.

1 Introduction

Interdisciplinary approach is integral to successful outcomes in research and practice in the information technology (IT) domain. This is mainly because of the highly interconnected nature of IT with business. As is evident in daily usage, basic business processes related to banking, travel and hospital are so meshed up with IT that they don't have an independent, manual existence. The upcoming Big data technology provides an excellent example of this intertwined nature of IT with business processes. Incorporating Big data in business requires an understanding of the many non-technical facets of the business—including economics (e.g. calculating ROI), sociology (e.g. project teams and their behavior), psychology (e.g. individual motivators and biases) and so on. In this paper we present how such interdisciplinary approach was used in the development of the Big Data Framework

B. Unhelkar (✉)
University of South Florida, Sarasota-Manatee Campus, Sarasota, FL, USA
e-mail: bhuvan.unhelkar@gmail.com; bunhelkar@sar.usf.edu

S.C. Suh, T. Anthony (eds.), *Big Data and Visual Analytics*,
https://doi.org/10.1007/978-3-319-63917-8_5

for Agile Business (BDFAB). This paper further argues why the implementation of this framework in practice also benefits with the use of an interdisciplinary approach.

2 Positioning the Need for Big Data and Agile Framework

Figure 1 outlines the philosophy for interdisciplinary research and application. To understand a problem it needs to be broken down into small, comprehensible elements. This is called 'Analysis'. However, once the problem is understood, then as a solution is designed and applied in practice. Applying a solution to a problem requires a holistic approach. This is called 'Synthesis'. Awareness of multiple disciplines is important in both analysis and synthesis.

The "Big Data Framework for Agile Business" (BDFAB) is a research-based framework that facilitates a strategic approach to application of Big Data to business. The development and use of this framework is based on multiple disciplines including technology, business, psychology, social science and finance. While most contemporary Big Data approaches focus either on the Hadoop eco-system (as a suite of technologies, programming and management) or on the Analytics (based around extensive statistical techniques such as predictive analytics, net promoter score—NPS—and so on), the BDFAB reported here takes a *strategic* approach to the use of Big data. Transcending technology and analytics to move into the strategic business space with Big Data is shown in Fig. 2.

Taking a strategic view to both Big data and the ensuing Agile business requires this framework to be *holistic*. And, a holistic framework needs knowledge, investigations, experimentation and practical experiences from widely dispersed

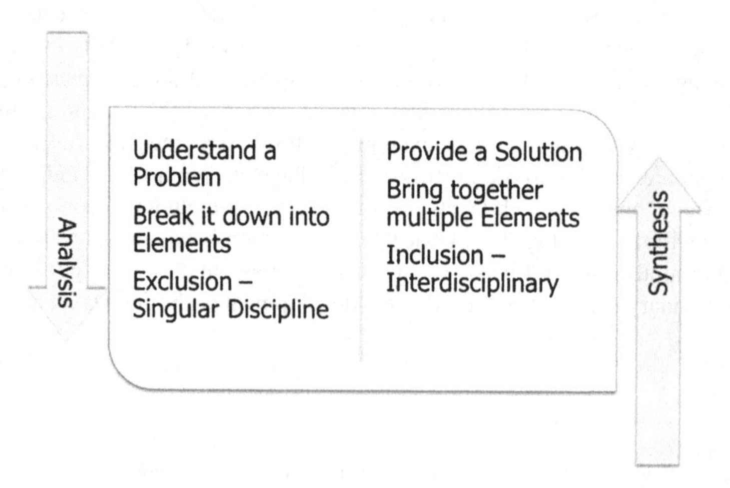

Fig. 1 Analysis vs. synthesis—basis for Interdisciplinary innovations, research and practice

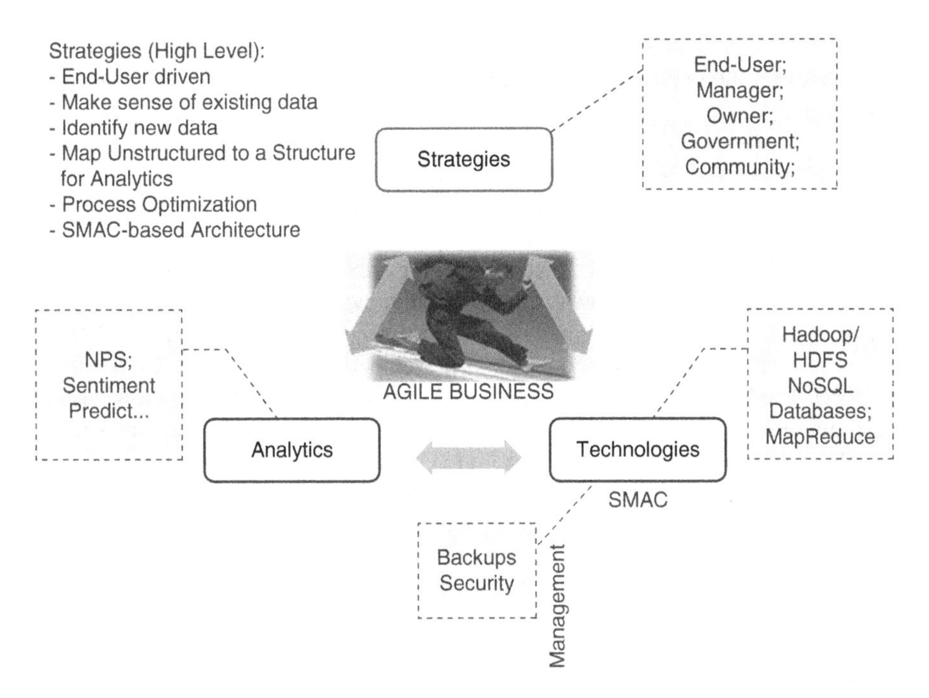

Fig. 2 Positioning big data strategies: transcending analytics and technologies © MethodScience

disciplines. Inclusion of the many social science disciplines makes the framework richer and adoptable. In fact, ICT is the cause for completely new business processes and business models. These new business processes are themselves characterized by the key word 'Agile.' Unhelkar [1] has argued in detail for the need for Agile to be considered as a business strategy—resulting in *Composite Agile Method and Strategy (CAMS)* [1, 2].

3 Big Data and Agile Business: Literature Review

The origin and basis for Agile as a method is encapsulated in the popular Agile Manifesto ([3]; also Fowler and Highsmith [4]). This manifesto characterizes Agile as "a value statement not a concrete plan or process" [5]—thereby laying the foundation for the Agile values, priorities and principles. The translation of these values from the depths of software development to business processes is the result of combining the formality of planned approaches and the flexibility of Agile–Boehm [6] has similarly argued: "Although many of their advocates consider the agile and plan-driven software development methods polar opposites, synthesizing the two can provide developers with a comprehensive spectrum of tools and options." The proximity of Agility with business has led to many practitioners exploring the synergy between Lean and Agile. This is a business-focused approach to Agile that capitalizes on the opportunities for rapid decision making to render lean business processes [7].

Agile in projects incorporates multiple disciplines. However, the role of psycho-social aspects in software projects, in particular, was very much in vogue. For example, in a special issue of *Communications of the ACM* commemorating the first 50 years of computing, virtual reality pioneer Jaron Lanier writes: "... The biggest surprise from the first 50 years of computers is that computation turns out to be a cultural object in its own right, with all its warts and semicolons." [8] This phenomenal importance of "human issues" in IT project management also finds way in discussions by DeMarco and Lister [9], Weinberg [10] and Larry Constantine [11, 12]. The latter states: "Good software does not come from CASE tools, visual programming, rapid prototyping, or object technology. Good software comes from people as does bad software." Unhelkar [13] has also discussed this importance of people in software projects and the destructive nature of "game playing"—together with suggested antidotes. Levison [14] takes this human side of projects further by providing its interesting insights in the context of Agile. Arguing that dealing with the human element is a key part of a successful agile transition. These discussions provide the basis for considerations of psychological, social and cultural factors in creating a practically applicable strategic framework. As mentioned earlier, this is particularly true in the domain of Big data. As Evernden [15] mentions, "As the volume of data grows, the role of information architecture is changing, from the passive structuring and managing of data to a smarter, more active role of information effectiveness."

The business strategy space is laced with the ever so important aspects of Agility. Hence, the way to enable businesses to tap into these opportunities (which range from expansion into new markets, enhancing customer satisfaction and/or achieving excellence in optimizing internal business processes) is by bringing in vital elements of Agile values, principles and practices as applied to business. BDFAB positions Agile in this strategic business context. The assertion here is that Agile has transcended software development and now plays a major role in business organizations [16]. Agile is therefore a legit business goal in its own right [1], and a strategic approach to Big Data can help a long way in achieving that goal. Such an approach aims to make use of structured, semi-structured and un-structured data, its velocity and its volume to generate ongoing and significant amount of Business Intelligence that would enable improved business decision making. In addition to providing the necessary support in terms of organization of Big Data through a reference framework, BDFAB also supports its adoption at an organizational level through a well-defined 12-lane adoption process.

Lampitt [17] presents how 'Relentless data growth has conjured up new technologies for data analysis and unprecedented opportunities for insight'. Lampitt further corroborates Gartner's definition of Big Data as "high-volume, high-velocity, and/or high-variety information assets that require new forms of processing to enable enhanced decision making, insight discovery and process optimization."

Kellen [18] further has this caveat: "Big Data will lead to big denial. The endless yin and yang of learning and denial that is probably forever part of the human condition ought to create some big opportunities for enterprising folks who can help us tame our denial instinct."

Shah et al. [19] mentioned that good data is not enough to ensure good decisions. This can be extended further to say that good analytics are also not enough for good decision making. BDFAB is based on the premise that good data and good analytics are meant to support core business; and that a much more strategic approach to Big data is imperative—that will lead into designing good analytics and ensuring utilization of good data. The starting point is the business *as is*, and the vision, *to be*. And that *to be* includes Agility—ability to respond rapidly and effectively to changing internal and external circumstances.

LaValle et al. [20] further underscore the importance of utilizing the insights generated through Big data analytics to provide *value*. This value is for the customers, the business partners and also internally for the organization's management. It is the importance of value of big data for business that brings in elements of inter-disciplines—psychology and sociology in particular—in this discussion.

Finally, as Jeff Jonas writes in his foreword to Dr. A. Sathi's book—*Big Data Analytics*, [21]) and observe how it maps to Agility in business:

- Organizations must be able to *sense* and *respond* to transactions in real time (Agility is the ability to be able to spot the changes coming through—which are transactions as both micro and macro levels)
- Also must be able to *deeply reflect* over what has been observed—to discover relevant weak signal and emerging patterns (Agility requires ability to take effective decisions; this effectiveness results from deep reflections, aided and impacted by Big data analytics).
- As this feedback loop gets faster and tighter, it significantly enhances the discovery (Agility requires rapid response which, in turn, is based on analytical insights and leanness of organizational structure)

4 BDFAB: Overview of the Framework

Based on the above literature review and practical experiences, the BDFAB has emerged as a comprehensive and holistic framework for application in business. The key elements of this framework and the corresponding fundamentals of those elements are summarized in Table 1.

BDFAB starts with the business organization itself—its SWOT analysis. This analysis is in the context of the vision and the necessary capabilities required of the organization to satisfy that vision. Helping the business identify and exploit the existing and growing Data capabilities (Technologies and Analytics) with a continuous focus on business decision making produces relevant insights. These are the Building Blocks which form the key part of BDFAB, and shown in the

Table 1 Overview of BDFAB—elements and business considerations

Elements of framework	Fundamentals (Examples only)	Business consideration
Values	Agility, insights, collaborations	What does the business aspire for? To_Be state
Roles (people)	Data scientist, user, analyst, coach, investor	Who are the people—to make it happen, to benefit?
Building blocks (phases)	Business decision (SWOT), technology, user experience, quality, people	Why do it? (Business reasons); How to do it? (phases)
Artefacts	Plans (Financial, ROI), feedback, approach, staff, centre of excellence	What to produce? To use?
Conditions	Type, size of business (as is)	Where and when to apply BDFAB?
Practices (agile)	Standups, stories, showcase	How do undertake Agility at change level
Compendium (repository)	Manifesto, strategy cube, adoption process	Guiding change management; transformation

third layer of the framework. The output of this building block enables decision makers to set new products and services, respond to individual customer issues and rapidly change the business processes. BDFAB helps utilize the technologies of Hadoop/MapReduce/Spark to directly impact the Agility of business. The decision makers need to see the correlation between Big Data (and its Analytics) and making the Business 'Agile' as a result of those analytics.

Technologies of Big Data are based around the Hadoop eco-system that includes the HDFS-based NoSQL databases and the MapReduce-based programming languages. These technologies are equipped to deal with vast amount of data—their storage, sharing and processing. They also handle unstructured data. High-level understanding of the nature of this data is an interdisciplinary exercise as it requires certain estimations and assumptions on the part of the Data scientist.

Analytics, on the other hand, focus on applying statistical techniques to this large amount of data in order to generate insights. Davenport [22] expands on the competitive power on analytics. Conducting the actual analytical work, therefore, is mono-disciplined in the sense that is focuses only on statistical knowledge and output—without worrying about how it will be applied in practice. Analytical techniques require both statistical and programming (e.g. 'R') knowledge and experience.

However, the analytics themselves are no longer static; they are themselves changing depending on the circumstances of an individual customer and/or the context in which the business finds itself (e.g. political uncertainty, changing legal structure, global collaborations). This requires Agility to be applied to the analytic processes themselves. The comprehensive framework for Big Data adoption (particularly in large organizations) comprises multiple disciplines which uplifts the capabilities of its decision makers on an ongoing basis resulting in business Agility.

Table 2 The six aspects of an Enterprise Architecture and corresponding interdisciplinary focus

Enterprise architecture		Disciplines
What	Technical	Technology—Hadoop Eco-System, Spark; 3xV of Data; Analytical—need to break down the problem into small elements
Why	Business	Economy (Financial)—ROI, Risks in Big Data adoption; Requires in-depth calculations, keeping the end-goal in mind
When	Project	Management; Scheduling Transformation; Requires detailed project planning to start with—but once the project comments requires ongoing holistic management
How	Analytical	Business Processes; embedding Analytics; Agility in business; Brings in the process discipline
Where	Project	Management; Administration; Location-specific or Location-independent awareness of the solution
Who	People	Sociology; Psychology; Quality (using SFIA to uplift Capabilities)

Table 2 shows the six aspects of an Enterprise Architecture.

The key innovation in terms of BDFAB as a framework for Big Data Strategies. This helps focus on business strategies as a vital starting point for utilization of Big Data domain. An accompanying and equally important aspect of this framework is the focus on Agile. This incorporation of Agile in BDFAB is based on the premise that Agile has transcended software development and now plays a major role in processes associated with the business. Agile business is the goal—and a strategic approach to Big Data can help a long way in achieving that goal.

5 BDFAB Modules (Building Blocks) and Their Interdisciplinary Mapping

While analytics including OLAP cubes, text and data mining and dashboards all add to and aid in decision making (see [23]), what is even more interesting is the strategy to put this whole process together. How does one get an organization to reach a stage where these analytics and their ensuing decision making becomes a norm? To achieve that goal, BDFAB comprises five major *modules* or Building blocks: These are listed in the third layer of the framework and are also summarized in Table 3 below.

Table 3 The five major modules (building blocks) of BDFAB 1.0 and their interdisciplinary mappings

BDFAB building blocks	Initial focus	Interdisciplinary aspects
Business decisions	Synthesis	Focuses on existing capabilities, future vision and a SWOT analysis
Data science: analytics, context and technologies	Analysis	Understands and builds on the technical capabilities of the Hadoop Eco-system; and the VVV of Big Data
Business processes: fine granular decision making	Synthesis	Builds on the Value theme; Analytics range from before the user comes in contact with the business to well after that
Enterprise Architecture—SMAC—TESP—Semantic	Synthesis	Technical, Economic, Social and Process dimensions of a business are affected by Big data and examined here. Social-Mobile-Analytics-Cloud (SMAC-stack) also examined here
Quality—GRC—People (Skills)	Synthesis	Use of SFIA framework to uplift the capabilities of people/individuals of an organization in the context of Big Data

5.1 Business Decisions

This module starts with the introduction of the concept of Big Data and positioning it for business strategies. The discussion here starts with the terminology and family structure of Big Data Analytics. While alluding to the technologies of Hadoop/MapReduce, the module is still focused on the strategic/business value of Big Data Analytics. Thus, we also discuss the various business approaches—starting with a SWOT analysis and moving into the risks, cost advantages and adoption approaches to big data. Furthermore, Agility as a business concept (transcending Agile used in software development projects) makes this a more holistic (synthesis) starting point.

5.2 Data Science: Analytics, Context and Technologies

The second module focuses on Data Analytics—Mapping Volume, Variety and Velocity with Structured, Unstructured and Semi-Structured data types. Each of these characteristics of Big Data is invaluable in supporting corresponding business strategies—if properly formulated. This module demonstrates the interplay between Analysis of Data and its impact on creating business strategies. Analysis takes the user into further discussion on the advantages as well as the challenges of analyzing unstructured data (by mapping/indexing it with a superimposed structure). Fundamentals of NoSQL databases and how they can be used for business advantages are highlighted here.

5.3 User Experience: Operational Excellence

This module explains how Data Analytics can render a business Agile. Understanding customer (user) sentiments through a User Experience Analysis Framework (UXAF) is the starting point for this work. Most UXAF focus on time T0 to time T1—when the user is in direct contact with the business through its systems and interfaces. Substantial data is generated, however, by the interactions of the user with his/her social and mobile networks that occur before T0 and after T1. Exploring the generation and use of this data (based around SMAC-stack) is part of the discussion in this module. The 'predictive' and 'prescriptive' nature of ensuing analytics is discussed here. Since UX is usually subjective, a synthesis of the issue is a good starting point.

5.4 Quality Dimensions and SMAC-Stack

Quality considerations in the Big Data domain assume prominence because of the direct impact they have on Business Decision making. This module focuses on this crucial Quality aspect in Big Data solutions: Data, Information, Analytics (Intelligence), Processes, Usability and Reliability. Uniqueness of un-structured data and what can be done to enhance and validate its quality are part of this discussion. Challenges of Contemporary Testing (and the role of Agile practices, such as continuous testing) together with their application to Big Data (Each Analytics needs immediate—Agile-like—Testing) is also explained here.

5.5 People (Capability)

Moss and Adelman [24] have discussed the ever growing importance of people and their capabilities in the Big data space. Similarly, the McKinsey [25] report on Big data goes into the details of existing and needed capabilities in the Big data technologies and analytics domains. This framework focuses on this important people issue—identifying and enhancing the capabilities at both technical and analytical level—using the Skills Framework for Information Age (SFIA).

6 Conclusions and Future Direction

This is a practical paper that reports on the development of a framework (BDFAB) that is unique in its applicability to business. This uniqueness comes from the interdisciplinary approach in developing and applying the framework. Furthermore, this framework also embeds aspects of psychology and sociology to ensure the solutions are applied in a holistic manner to business. BDFAB tends to be unique in the sense that it elevates the current industry focus from technologies and analytics to business strategies. Another important part of BDFAB is its coverage on resourcing (up-skilling, training, recruitment and coaching—derived from the application of Agility) and people-focus. Most big data approaches consider people/resourcing as an afterthought whereas BDFAB draws attention to this vital element upfront—resulting in formation of Centres of Excellence around Big Data and related disciplines. Future directions include accompanying BDFAB with a corresponding CASE tool and metrics.

Acknowledgment University of South Florida, Sarasota-Manatee campus; MethodScience.

References

1. Unhelkar, B.: The Art of Agile Practice: A Composite Approach for Projects and Organizations. CRC/Auerbach, New York/Boca Raton (2013)
2. Mistry, N., Unhelkar, B.: Composite agile method and strategy: a balancing act. Presented at the Agile Testing Leadership Conference 2015 – held 21st Aug, 2015 in Sydney, Australia; (Organized by Unicom) (2015)
3. Agile Manifesto for Agile Software Development: viewed 9 March 2009, http://www.agilemanifesto.org/ (2001)
4. Fowler, M., Highsmith, J.: The agile manifesto, Tech Web, viewed 9 March 2009, http://www.ddj.com/architect/184414755 (2001)
5. Coffin, R., Lane, D.: A practical guide to seven agile methodologies, part 1, Jupitermedia Corporation, viewed 29 September 2009, http://www.devx.com/architect/Article/32761/1954 (2007)
6. Boehm, B.: Get ready for agile methods, with care. IEEE Computer. **35**(1), 64–69 (2002)
7. Unhelkar, B.: Lean-agile tautology. Cutter, Arlington (2014)
8. Lanier, J.: The frontier between us. Commun. ACM. **40**(2), 55–56 (1997)
9. DeMarco, T., Lister, T.: Peopleware: Productive Projects and Team. Dorset House, New York (1987)
10. Weinberg, G.M.: The Psychology of Computer Programming. Van Nostrand Reinhold, New York (1971)
11. Constantine, L.: Panel on "soft issues and other hard problems in software development." In: OOPSLA'96, San Jose, California, USA, October 1996 (1996)
12. Constantine, L.: Constantine on Peopleware. Yourdon, Raleigh, NC (1995)
13. Unhelkar, B.: Games IT people play, information age, publication of the Australian Computer Society, June/July, 2003, pp. 25–29 (2003)
14. Levison, M.: "Why agile fails at scale: the human side" Agile Product & Project Management, Cutter E-Mail Advisors, USA 15th Dec, 2011 (2011)
15. Evernden, R.: Information architecture: dealing with too much data, Cutter executive report, Boston, USA (2012)
16. Unhelkar, B.: Agile in practice: a composite approach. Cutter Consortium Agile Product & Project Management Executive Report, vol. 11, no. 1, Boston, USA (2010)
17. Lampitt, A.: Relentless data growth : the big data journey has begun. In: InfoWorld (2012). https://www.infoworld.com/article/2607256/big-data/the-big-data-journey-has-begun.html. Accessed 7 Oct 2017

18. Kellen, V.: Cutter E-mail advisory. https://www.cutter.com/article/big-data-big-denial-377591 (2013). Accessed 18 April 2013
19. Shah, S., Horne, A., Capellá, J.: Good data won't guarantee good decisions. Harv. Bus. Rev. **90**(4), (2012). https://hbr.org/2012/04/good-data-wont-guarantee-good-decisions
20. LaValle, S., Lesser, E., Shockley, R., Hopkins, M.S., Kruschwitz, N.: Big data, analytics and the path from insights to value. MIT Sloan Manag. Rev. **52**(2, Winter), 21 (2011)
21. Sathi, A.: Big data analytics: disruptive technologies for changing the game, MC Press Online, LLC, Boise ID 83703, USA. © IBM Corporation (2012)
22. Davenport, T.H.: Competing on analytics. Harv. Bus. Rev. 99–107 (2006). https://hbr.org/2006/01/competing-on-analytics
23. Kudyba, S.: Big Data, Mining, and Analytics: Components of Strategic Decision Making. CRC/Auerbach, New York/Boca Raton (2014)
24. Moss, L.T., Adelman, S.: The role of chief data officer in the 21st century. Cutter exec report, vol. 13. Cutter Consortium, Boston (2013)
25. McKinsey report on Big Data.: http://www.mckinsey.com/business-functions/digital-mckinsey/our-insights/big-data-the-next-frontier-for-innovation viewed on 30th April, 2017 (2011)

Scalable Gene Sequence Analysis on Spark

Muthahar Syed, Taehyun Hwang, and Jinoh Kim

Abstract Scientific advances in technology have helped in digitizing genetic information, which resulted in the generation of the humongous amount of genetic sequences, and analysis of such large-scale sequencing data is the primary concern. This chapter introduces a scalable genome sequence analysis system, which makes use of parallel computing features of Apache Spark and its relational processing module called Spark Structured Query Language (Spark SQL). The Spark framework provides an efficient data reuse feature by holding the data in memory, increasing performance substantially. The introduced system also provides a web-based interface, by which users can specify the search criteria, and Spark SQL performs search operations on the data stored in memory. Experiments detailed in this chapter make use of publicly available 1000 genome Variant Calling Format (VCF) data (Size 1.2TB) as input. The input data are analyzed using Spark and the end results are evaluated to measure the scalability and performance of the system.

1 Introduction

Big data computing technologies have been proliferating fast with unprecedented volumes of data generation in several domains. Since the completion of the Human Genome Project in 2003 [1], the cost of sequencing human genomes is reduced by more than 100,000× [2]. Reduced cost and recent scientific advances in technology have helped digitize human genes and generate huge amounts of human genomic sequences that help researchers in understanding human genes. A major challenge encountered with such vast gene data generation is the analysis of the required sequence structures.

Many data processing techniques are available, but the ability to process the increasing data needs to address the performance and scalability challenges. For

M. Syed • J. Kim (✉)
Texas A&M University, Commerce, TX, USA
e-mail: muthahar.msd@outlook.com; jinoh.kim@tamuc.edu

T. Hwang
University of Southwestern Medical Center, Dallas, TX, USA
e-mail: taehyun.cs@gmail.com

S.C. Suh, T. Anthony (eds.), *Big Data and Visual Analytics*,
https://doi.org/10.1007/978-3-319-63917-8_6

instance, processing of increasing data by minimizing the query response time is key to achieving scalability. One of the tools available for such large-scale analysis is Apache Hadoop [3], which is an open source framework that gives users powerful features to store and process huge datasets. While widely used, we observed it requires around 20 min for every single query to be completed against a data size of 1.2 terabytes (for the 1000 genome data set) using 16 Hadoop computing resources. This motivates us to explore other tools for scalable gene sequence analysis. Apache Spark [4] can provide enhanced performance for certain applications compared to the existing MapReduce framework using its in-memory processing feature. In detail, Spark provides Resilient Distributed Datasets (RDDs) that can be stored in memory between query executions that facilitates the reuse of intermediate outputs.

In this chapter, we examine the feasibility of Spark for the application of gene sequence variation analysis with respect to performance and scalability. In addition, optimization is an important part of system performance and we will evaluate the impact of caching in Spark with extensive experiments on performance evaluation. Another important element for such analysis systems is the degree of user friendliness to access the analysis tool since the majority of the users of this application would be scientific researchers with limited knowledge on computing systems. This chapter also introduces a graphical user interface for intuitive analysis and develops a prototype system that equips a variety of features to improve user friendliness including an automatic query generation engine.

Figure 1 shows an overview of the introduced system for large-scale gene sequence analysis with low latency for processing. The system uses the Spark framework with Spark SQL and YARN resource management that provides effective configuration of the cluster computing system. The graphical user interface of the system promotes user interaction as part of data analysis. As shown in Fig. 1, researchers can simply access the system interface on existing web browsers for composing a query. Based on the users' selection criteria, application system

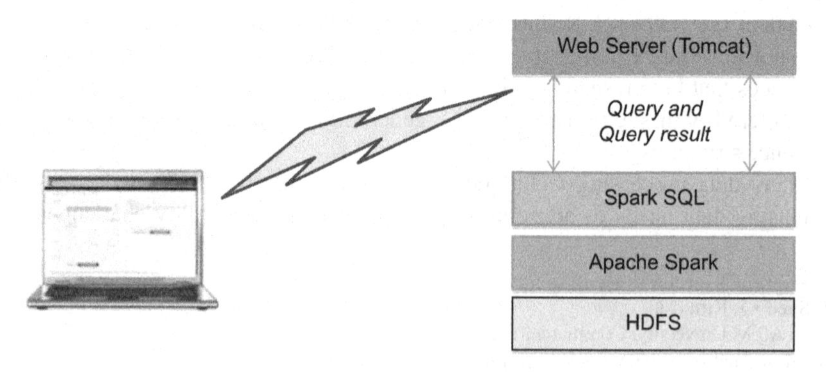

Fig. 1 System overview: The system is based on the Spark framework with Spark SQL and YARN resource management. End users access the system through Web-based interfaces

generates a query script automatically and executes the query script on Spark. Finally, the processing results are delivered to the end users.

This chapter is organized as follows. Section 2 provides the summary of the big data computing tools including Hadoop and Spark, with closely related studies to our work. Section 3 demonstrates the design of the system for scalable gene sequence processing on the Spark framework, and Sect. 4 presents the evaluation results conducted on a Spark computing cluster with 16 compute nodes. In Sect. 5, we present a prototyping system and we conclude our presentation in Sect. 6.

2 Background

2.1 Apache Hadoop and Spark

Hadoop is an open source software platform managed by the Apache Software Foundation. It is the most widely recognized platform for efficient and cost-effective storage and management of large volumes of data. HDFS provides a resilient and fault-tolerant storage of data by means of data duplication which is termed as data replication. Pig [5] and Hive [6] are the two high-level query languages supported by Hadoop, whose execution is based on MapReduce programming model that provides a scalable, flexible and fault-tolerant computing solution [7]. Hadoop comprises of two major components: Hadoop Distributed File System (HDFS) and the MapReduce framework. A traditional Hadoop job works by reading the data from HDFS for each run and writing back the intermediate results to HDFS which incurs the cost of data operations for each run time. Thus, Hadoop does not seem to be a good fit for iterative computations and low latency applications.

Spark is an open source framework that can process data 100× faster than Hadoop. It is a widely accepted for parallel cluster computing systems in both science and industry [8]. This framework attracts users due to the delivery speed it provides by making use of its in-memory computing capability using fault-tolerant [9] Resilient Distributed Datasets (RDDs).RDDs are created by performing operations called transformations (map, filter etc.) on the data stored in the file systems or other existing RDDs. By default, RDDs will be persisted in memory but they can even spill to the disk if there is not enough Random Access Memory (RAM). Spark also provides users with an option to specify the storage level for RDDs, such as persisting in disk only, disk by using replicating flag and both memory and disk. Spark provides Scala, Java, and Python programming interfaces to create RDDs transformations and performs actions on it [10]. Spark contains Spark SQL module that helps to achieve relational processing on the persisted data.

2.2 Gene Sequence Analysis

Genes are small pieces of a genome and they are made of DNAs. DNA contains bases that carry the genetic information which helps in understanding the behavioral pattern of organisms. Each human genome contains approximately three million DNA base pairs. Genome sequencing is about identifying the order of DNA bases that defines the characteristics of an organism. Genome sequences help researchers or scientists answer questions like why some people get infections or why some people have different behaviors that others do not have, and will the genes have any impact on others genes.

Many computational tools have been developed for digitizing the genetic information, producing genome sequences which can be further analyzed for meaningful patterns. As mentioned, one genome has up to three million base pairs and thus sequencing of multiple human genomes will quickly add up to larger data sizes and mapping, in turn, generates more. For effective analysis of such data, it is essential to store data in a fault-tolerant manner and process the data with optimal performance and scalability. To address this concern, Hadoop has been applied in the Genome Analysis Toolkit (GATK) tool [11]. In the tool, the GATK process needs to access the distributed data across the cluster, parallelize the processing, persist and share intermediate processing results. This process has drawbacks such as limited parallel processing with the significant overhead due to disk I/O between stages of walkers and expensive read while processing between stages for repeated access. The idea of caching in Spark could *reduce* the repeated disk access of data that eliminates the repeated I/O bottleneck and improves the performance.

Spark is used by variant calling tools for the genomic analysis. Using the advantages of Spark, the ADAM project was developed by Big Data Genomics (BDG) group at UC Berkeley. ADAM can process large-scale genome data available in various data formats such as VCF, BAM, and SAM [12]. ADAM project uses APIs that transform the input data to ADAM-defined formats and store the RDD sets generated from these formats in memory to perform sorting and other operations using Spark modules such as GraphX, MLlib (Machine Learning libraries) and Shark. Even though ADAM has overcome the traditional processing speed limitations of MapReduce, additional processing complexities such as converting VCF files to the ADAM formats and overhead for compression of inputs before processing, have led to non-optimal performance.

Another Spark based implementation for genomic information analysis is VariantSpark [13]. This system is developed in Scala using Spark and its machine learning methods (MLlib). VariantSpark reads the VCF files and transposes them into vectors based on variants, which are then transformed into key-value pairs. The key-value pairs are then zipped and saved in a distributed system. Further processing is implemented using a K-means clustering algorithm. MLlib performs complex programming logic by transforming, zipping and processing of data using machine learning algorithms.

Unlike the above Spark-based gene sequence analysis system, our system employs Spark SQL tables to process the direct genome VCF data without conversion and uses Spark in-memory data store for processing required data, in order to reduce format conversion and save processing times. The introduced system concentrates on reducing the processing complexities, maximizing the performance of the variants analysis, and increasing the ease of usability of the tool when compared to the other variant tools.

3 System Model for Scalable Gene Sequence Analysis

Utilizing the advantages of HDFS, Sparks' relational processing module Spark SQL and YARN cluster management, a scalable model is designed as shown in Fig. 2. A series of experiments were conducted on top of this design to examine the scalability. Spark applications can run on top of existing HDFS that store data in clusters in a resilient manner. There are three possible cluster modes that control the resources of a cluster, namely, Standalone Mode, Mesos Mode, and YARN mode. Standalone Mode utilizes all the recourses available in the cluster and at least, each executor has to be running on each node. Total utilization of resources causes other applications to be in the queue until the prior process is completed, which will be a downside when processing multiple applications. To overcome this, YARN resource manager can be used as it can allocate resources dynamically within the cluster or we can choose the number of executors to be assigned to each process [14].

In order to process the data in a cluster, Spark application runs as an independent process in each node of the cluster and these processes are coordinated by a driver program called SparkContext. As shown in Fig. 2, the driver program interacts with Resource Manager in YARN Mode, which keeps track of the cluster resources and monitors tasks allocated to Node Mangers. Node manager acts as a worker that initiates tasks based on allocated cores for processing of data stored in its respective nodes. Each node can have multiple executors and each executor can have multiple tasks. Each node also has designated cache memory to store the data which can be accessed by all the executors in the node.

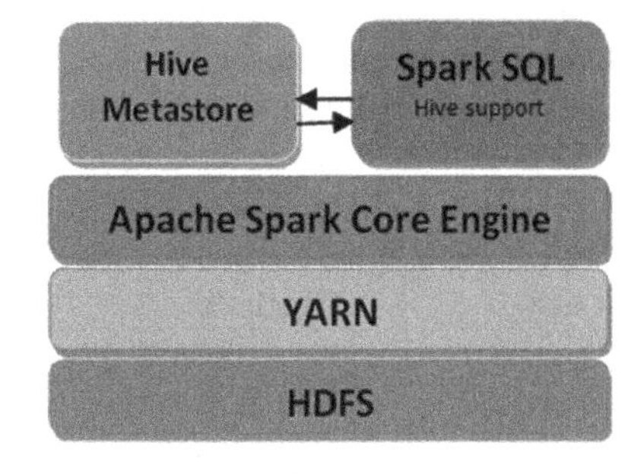

Fig. 2 System model, utilizing the advantages of HDFS, Sparks' relational processing module Spark SQL and YARN cluster management

Spark provides a Thrift JDBC/ODBC server that helps to run queries from within applications or by end-users using a command-line interface called Beeline. This acts as Business Integration medium in connecting different applications to Spark. This server provides distributed engine that can run SQL queries without writing complex code for querying. Beeline command lines are used to test the JDBC server connectivity from applications to run the SQL queries. This thrift server helps in maintaining the cache as long as the server is active. Keeping the server active helps multiple users access the cache to obtain faster results. This server is used in the design of proposed prototype system for efficient and optimized gene sequence analysis.

We assume that VCF [15] format genomic data is read and processed using Spark to identify if the system is scalable to use. In addition, a web-based gene Sequence analysis system is designed to store the VCF formatted data and process the data interactively for faster and easier access as will be discussed in Sect. 5.

4 Evaluation

In this section, we present the experimental results conducted particularly to evaluate the scalability of the system.

The experiments were conducted in a computing cluster machine that consists of 16 nodes mounted in a rack. Cluster configuration used for the experiments is of 128 cores, 128 GB memory and 32TB disk space. The nodes in the cluster are interconnected through Gigabit Ethernet Switch. The Operating System used for the cluster is CentOS release 6.5 that comes with rocks application, which manages the connections between all nodes that make up a cluster machine. I have installed the Hadoop pre-built version of Apache Spark v1.4.1, Stable version of Hadoop v2.6.0 and high-level query language Pig v0.14.0, Hive v1.2.0 and MySQL Server v5.6 for the experiments.

The experimental data set used has VCF formatted F data file [16], which contains genotype data of 1,092 individuals which equal to 37,644,734 records that approximate to 1.2TB. VCF represents Variant Calling Format that stores the genome data in the form of variants or subjects arranged as columns in the data file. The data are distributed to the fault tolerant HDFS across 16 nodes in the cluster. To consider the impact of query characteristics, three types of queries that have different coverage with respect to the fraction of the records hit are defined. Table 1 presents a summary of the queries with the description and hit coverage.

Table 1 Query definition

Query	Description	% hits
Query1/Q1	$1 \leq$ CHROME ≤ 5 and REF = ‘C’ and ALT = ‘G’	1.44%
Query2/Q2	CHROME = 1 and $1{,}000 \leq$ POS $\leq 200{,}000$	<0.01%
Query3/Q3	QUAL ≥ 100	94.80%

Table 2 Query performance for Spark SQL and other high-level query tools (unit: seconds)

	Pig	Hive	Spark (w/o cache)	Spark w/ Cache (first run)	Spark w/ cache (second run)
Query1	1260	1247	603	6.88	6.8
Query2	1258	1244	598	5.22	5.2
Query3	1290	1114	614	8.67	8.7

Query2 and Query3 are extreme cases with a very rare hit ("small hit") and a large hit ("large hit"), respectively. SQL aggregate functions executed on data have less output display time compared to the select functions on the data. Hence, the queries are chosen to perform the count operations on data to minimize the output display time and thereby to measure the computational performance exclusively.

4.1 *Spark SQL vs. Other High-Level Query Tools (Pig and Hive)*

For big data analytic tools, high-level query support is essential for easy usage and reducing programming complexities. Input dataset is loaded into HDFS, and high-level query languages are often used to process the loaded data. We compared several high-level query tools such as Pig and Hive with Spark SQL.

Table 2 shows the query completion time in seconds. The results show that Spark's iterative response time gives approximately 90–100% decreases the response, increasing the performance, when compared to Spark's initial response time (Cache Time) and approximately 200% increase when compared to Pig and Hive's response time.

4.2 *Scale-Out and Scale-Up Performance*

To check the scalability of using Spark for analysis the experiments are conducted in two folds. First, we evaluate scale-out performances of the proposed system. The scale-out check is to evaluate the impact of increasing computing capability for processing of data. Second, we evaluate scale-up performance of the proposed system. This scale-up check is to examine the impact of increasing data sizes for processing the data using the proposed system.

Figures 3 and 4 show scale-out performance in case of in-memory data access (Fig. 3) and in-disk data access (Fig. 4). The experiments are carried on the cluster using one executor and three cores on each node. The two figures show a certain degree of scalability. The benefit is the greatest from four nodes to eight nodes in both settings. As expected, Query 3 (large hit) shows the greatest improvement with an increasing number of compute nodes. When we compare the two settings, in-disk

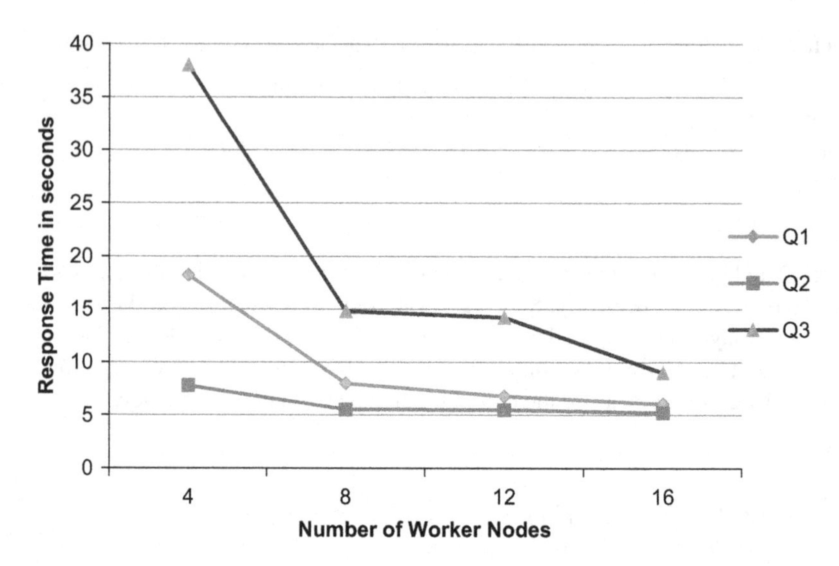

Fig. 3 Scale-out performance in case of in-memory data access

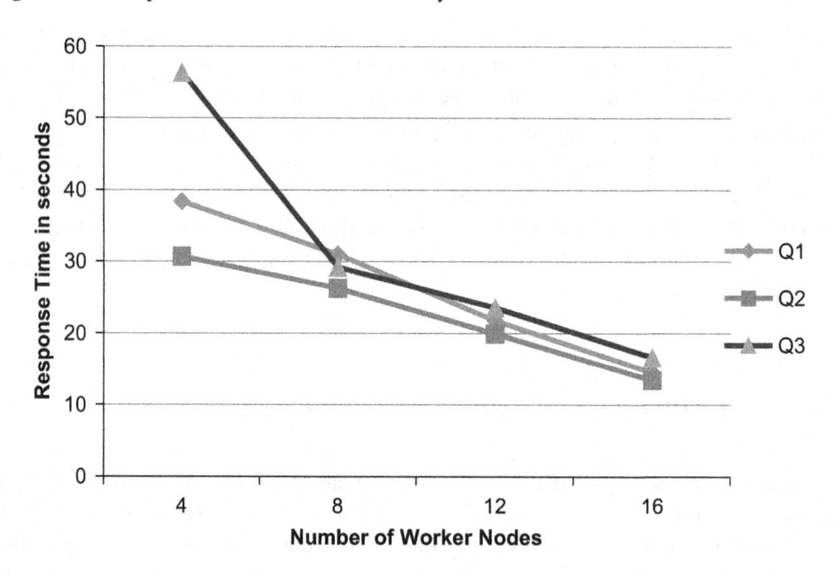

Fig. 4 Scale-out performance in case of in-disk data access

data access shows the greater scale-out performance. We presume that it is due to the basic overhead (around 5 s) to run an individual query.

We next discuss scale-up performance with increasing data sizes. The 1.2TB Genome dataset is divided into multiple datasets of increasing sizes from 10% to 100% of the total. For the Spark initial run or Cache run, an increase in data sizes results in greater processing time. Figure 5 shows the processing time that was taken

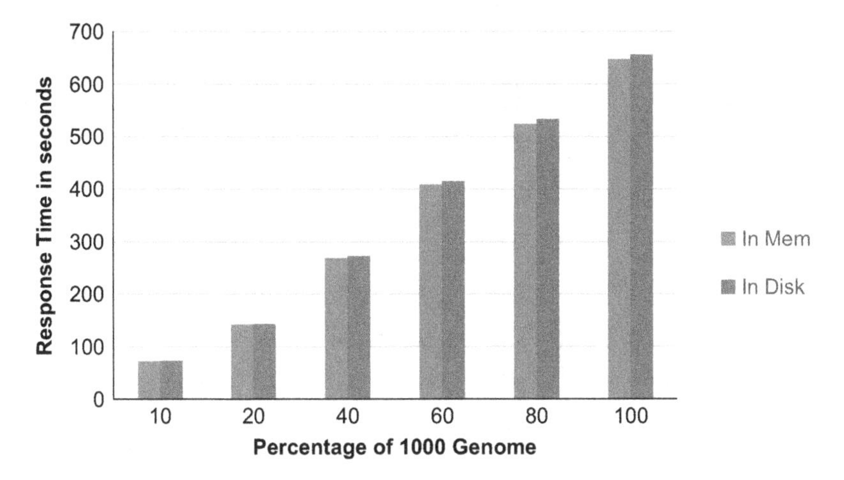

Fig. 5 Scale-up performance for cache loading time and initial run

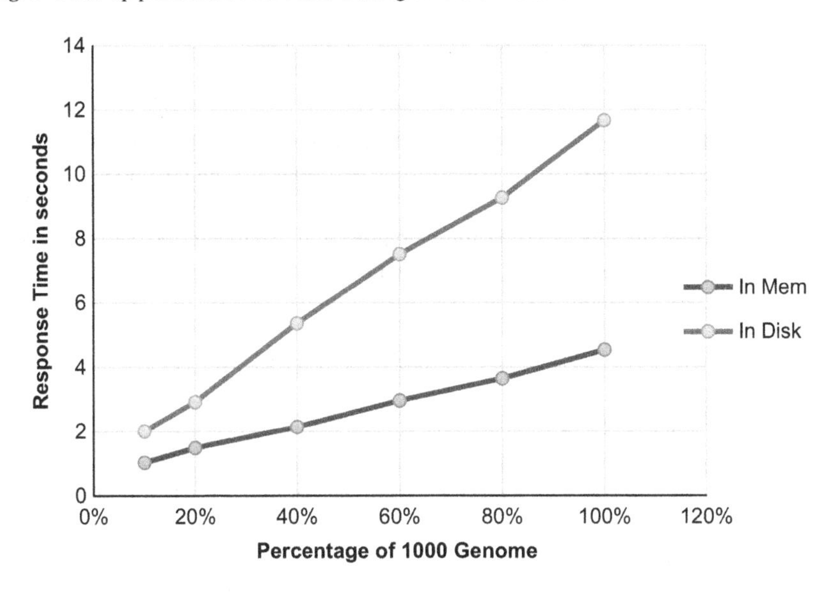

Fig. 6 Scale-up performance for cache scan

to cache the data in memory and on disk and increasing data sizes retain the linear scale-up performance of the system. Figure 6 shows scale up of the scan operation performed on the records cached in memory and on disk. After loading 10% percent genome, 16 columns of data is cached in memory or on disk for iterative runs. From both figures, we can see that initial loading time is tremendous; however after the data is loaded, we can see relatively very small scan time with a good degree of scalability.

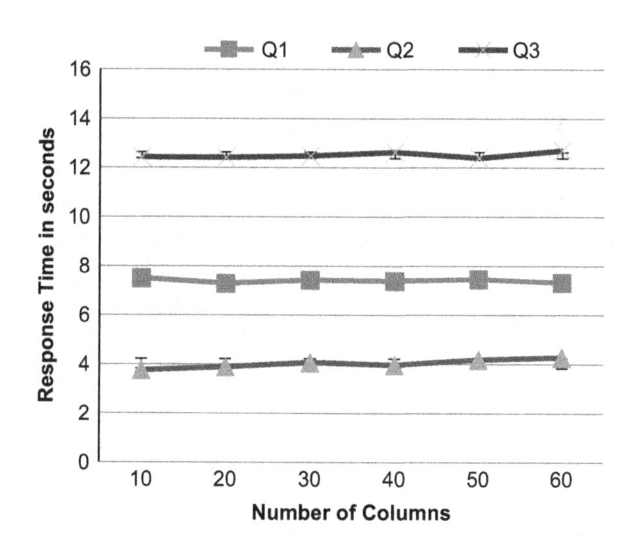

Fig. 7 Impact of number of columns cached

4.3 Impact of the Number of Cached Columns

We also evaluated the impact of the number of cached columns. As shown from Fig. 7, increasing the number of columns cached does not much impact the data response times. To test this, a series of query runs are implemented on the cached data by increasing the number of columns but with the same number of rows every time. Despite the increase in the amount of data loaded in cache, the processing time of queries on data remains consistent.

4.4 Impact of Data Spills

Spark caches the data in the memory based on configurations provided. If the memory available to cache is less than the amount of data to be cached, then Spark spills the leftover data after caching on to disk to avoid data loss. A series of queries are executed on cache data to test the impact of data spill from memory on performance. The configuration for this experiment is set to 16 executors with three cores and 1GB of memory allocated. From this configuration, the total memory capacity available for storing of data is up to 8.2GB.

As shown in Table 3, due to limited memory capacity, if users try to load more data into cache beyond the capacity that it can accommodate, then the remaining data will be spilled onto disk across each computing node within the cluster. The data spill on each node of the cluster is based on the size of data distributed across the nodes. From the previous experiment, we observe that each query executed on cached data with an increasing number of columns cached gives consistent processing time. However, an increasing number of columns cached will increase

Table 3 Response times in processing the spilled data from memory

Number of columns	Columns data size	Data cached in memory	Data spilled in disk	Query1	Query2	Query3
10	6.45GB	6.4GB	0.05GB	7.88	4.02	15.09
20	9.3GB	7.6GB	1.7GB	8.59	4.8	18.21
30	11.9GB	8GB	3.9GB	10.11	5.68	20.06
40	14.6GB	8.2GB	6.4GB	14.3	6.83	25.73
50	17.6GB	8.2GB	9.4GB	14.6	7.05	27.28
60	21.3GB	8.2GB	13.1GB	16.1	7.34	30.48

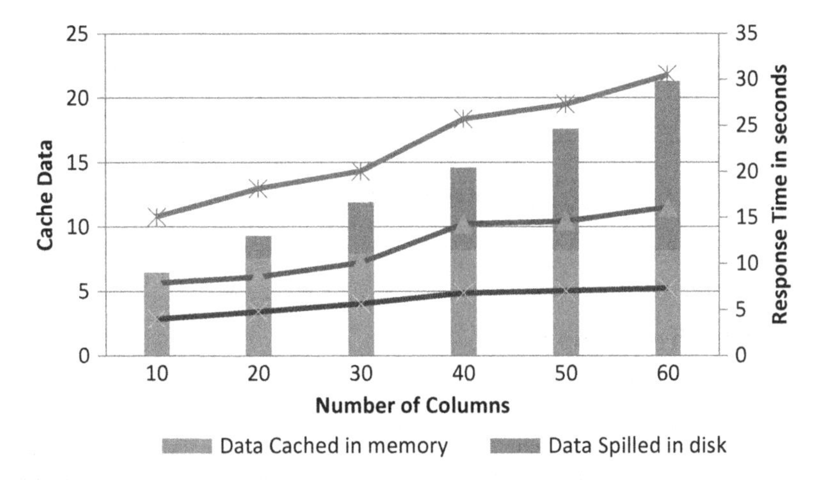

Fig. 8 Impact of data spills from memory

the size of data spill from memory due to limited capacity, which in turn impacts the processing time of the cached data. Figure 8 shows that increase in data spilled to disk increases the cached data processing time, and thus, decreasing the performance.

4.5 Impact of Record Hits

Accessing in memory data using different query options with varied output results will impact the data processing time. Within the cached data, increase in the number of records accessed (i.e. number of record hits) increases the processing time for each run. To demonstrate this, we loaded 16 columns of genome data in memory with 3.9 million records and performed a set of scan count queries. These queries hit the cache data with increasing percentage of output records. The results are plotted in Fig. 9, which shows that increasing access to the records from the cache data increases the processing time which ultimately leads to lower performance.

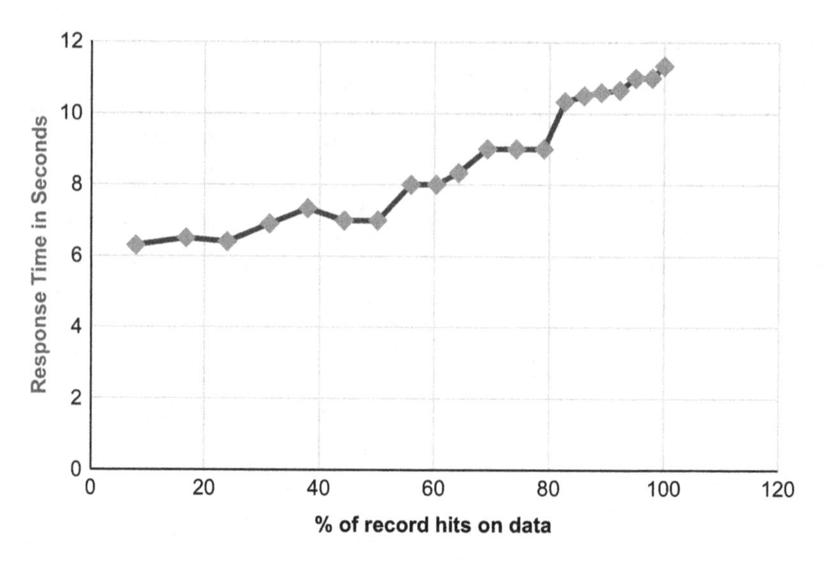

Fig. 9 Impact of record hits

5 System Prototyping

A prototype system is implemented to analyze the genome data using Spark. Figure 10 shows the proposed system architecture for gene sequence analysis that is designed to read the input VCF genome data and analyze the data using query scripts.

A web-based interface system designed to store and processes the VCF data in a more user-friendly manner. Figure 11 gives the flow diagram of this interface based on which the web interface is designed. Figure 12 displays the GUI interface designed for providing the query support to users, especially scientists with minimal programming skills.

This prototype system provides an option to execute queries in count only mode to maximize the performances. It minimizes the time that is taken to transfer the queried results between execution engine and the graphical user interface compared to transfer time without count mode. Larger the queried results the more will be transferred time of results.

Based on the above criteria editable query will be generated on the right-hand side of the interface for execution, as shown in Fig. 12. The generated query is executed and the output of executed query displays results along with record count and run time taken by Spark, as shown in Fig. 13. The query runtime does not include the data transfer time between Spark engine and the user interface module.

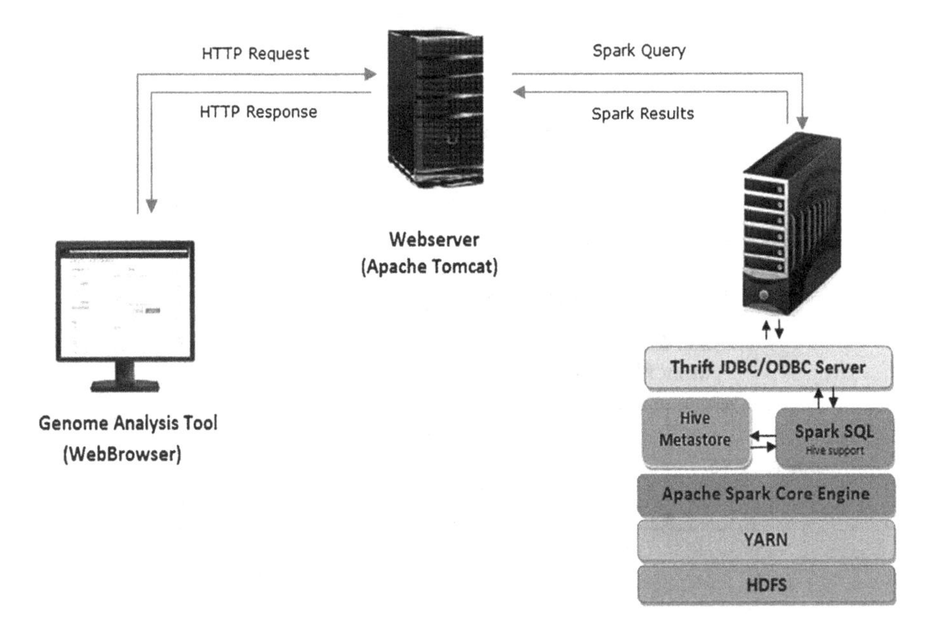

Fig. 10 Gene sequence analysis system architecture

6 Conclusions

Big data computing and analytics open the doors for researchers to play with an unprecedentedly large set of data in a cost-effective manner with efficient parallel computing clusters. This chapter makes use of these advanced analytic tools to develop a scalable system for gene sequence variation analysis. The key summary of this work are as follows:

- In this work, we examined the feasibility of using Spark for our own application with respect to performance and scalability, with the comparison with other analytics tools of Pig and Hive. We observed that Spark SQL significantly outperforms Pig and Hive. After loading the data in memory, it takes no longer than 15 s by Spark to complete any type of queries (execution plus response time) in our experimental setting with 16 computing nodes, while the other two tools require greater than 18 min for any individual query. The architecture used in this system with Spark and its ecosystem also demonstrated a high degree of scalability, particularly for scale-up over data size, suggesting it as a good choice for the bioinformatics application.
- We conducted extensive experiments to analyze ways to achieve optimal performance of the system. Even though iterative operations on the cached data will give results in seconds, the maximum number of active tasks or cores allocated for performing of cached data results in further minimizing the latency of jobs. Caching of any number subjects does not impact the processing time of queries on the cached data. Extended caching of data beyond the available cache memory degrades the performance of the system.

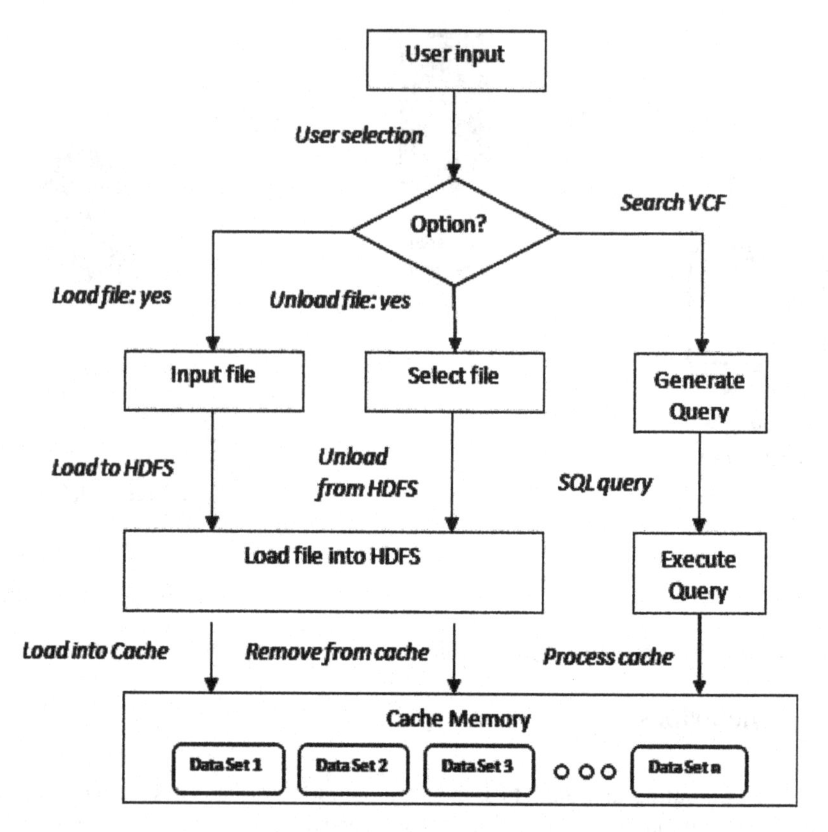

Fig. 11 Execution flows of the prototype system

- We implemented a prototype system that can process VCF files of any number of subjects. In this prototype system, users will load the data into the cache, after which the system generates and executes queries for the selected criteria's giving optimal results. The data will remain in the cache as long as user explicitly removes the cache. Hence, the holding of data in cache for longer duration helps users perform various query operations on data iteratively.

The prototype system can be further enhanced to process all formats of genome data along with VCF. In addition, providing a function for customization to individual users would be helpful for more flexible analysis. For example, users may want to configuration certain properties, such as the number of subjects to be cached, the size of the cache memory to be used, the size of the cache memory that is to be made available, and so forth. This work leaves such functions for greater flexibility for composing queries as one of the future studies.

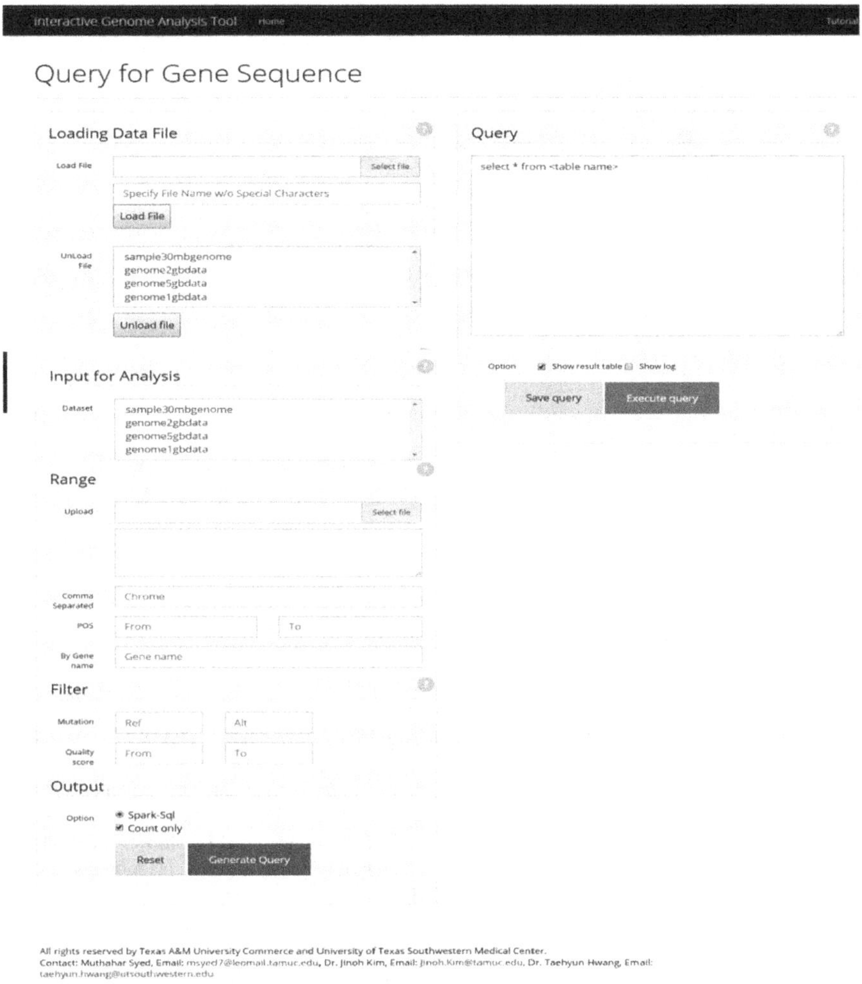

Fig. 12 Web-based user interfaces for query support

Since the genome data contains sensitive information of an individual, it is important to protect the data from unauthorized access. To ensure the privacy of the data, the system can be enhanced to implement encryption or privacy preserved algorithms and evaluate if there is any performance trade-off between processing data before encryption and after encryption. The presented system in this chapter does not contain a technical module for privacy and it will be an interesting piece of the future research.

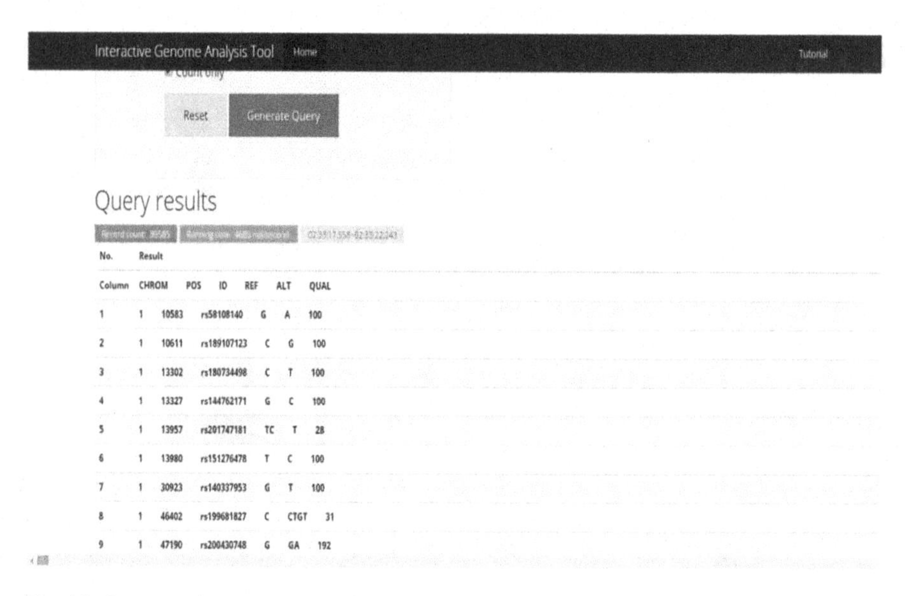

Fig. 13 Query result

References

1. Human Genome Project: [Online]. Available: https://www.genome.gov/10001772. Accessed 22 Nov 2015 (2003)
2. DNA Sequencing Costs: [Online]. Available: http://www.genome.gov/sequencingcosts/. Accessed 31 Jan 2016 (2016)
3. Lewis, S., Csordas, A., Killcoyne, S., Hermjakob, H., Hoopmann, M., Moritz, R., Deutsch, E., Boyle, J.: Hydra: a scalable proteomic search engine which utilizes the Hadoop distributed computing framework. BMC Bioinf., **13**, 6 (2012)
4. Apache Spark Project: [Online]. Available: http://spark.apache.org. Accessed 2015 (2015)
5. Apache Pig: [Online]. Available: http://pig.apache.org/ (2015)
6. Apache Hive: [Online]. Available: https://cwiki.apache.org/confluence/display/Hive/Home (2016)
7. Weil, S.A., Brandt, S.A., Miller, E.L., Long, D.D.E., Maltzahn, C.: Ceph: a scalable, high-performance distributed file system. In: Proceedings of the 7th Symposium on Operating Systems Design and Implementation (2006)
8. Zaharia, M., Chowdhury, M., Franklin, M.J., Shenker, S., Stoica, I.: Spark: cluster computing with working sets. In: Proceedings of the 2nd short Title USENIX Conference on Hot Topics in Cloud Computing, HotCloud'10, Berkeley, CA, USA (2010)
9. Zaharia, M., Chowdhury, M., Das, T., Dave, A., Ma, J., McCauley, M., Franklin, M.J., Shenker, S., Stoica, I.: Resilient distributed datasets: a fault-tolerant abstraction for in-memory cluster computing. UCB/EECS (2011)
10. Apache Spark Documentation: [Online]. Available: http://spark.apache.org (2016)
11. McKenna, A., Hanna, M., Banks, E., Sivachenko, A., Cibulskis, K., Kernytsky, A., Garimella, K., Altshuler, D., Gabriel, S., Daly, M., DePristo, M.A.: The Genome Analysis Toolkit: a MapReduce framework for analyzing next-generation DNA sequencing data. Genome Res. **20**, 1297–1303 (2010)

12. Massie, M., Nothaft, F.A., Hartl, C., Kozanitis, C., Schumacher, A., Joseph, A.D., Patterson, D.: ADAM: Genomics Formats and Processing Patterns for Cloud Scale Computing. DEECS Department, University of California, Berkeley (2013)
13. O'Brien, A.R., Saunders, N.F.W., Guo, Y., Buske, F.A., Scott, R.J., Bauer, D.C.: VariantSpark: population scale clustering of genotype information. BMC Genomics. **16**(1), 1052 (2015)
14. White, T.: Hadoop: The Definitive Guide. O'Reilly Media, Newton, MA (2015)
15. VCF File processing: [Online]. Available: http://vcftools.sourceforge.net (2014)
16. 1000 Genome Data: [Online]. Available: http://www.1000genomes.org/data (2014)

Big Sensor Data Acquisition and Archiving with Compression

Dongeun Lee

Abstract Machine-generated data such as sensor data now comprise major portion of available information, which raises two important problems: efficient acquisition of sensor data and the storage of massive sensor data collection. These data sources generate so much data quickly that data compression is essential to reduce storage requirement or transmission capacity of devices. This work first discusses a low complexity sensing framework which enables to reduce computation and communication overheads of devices without much compromising the accuracy of sensor readings. Then a new class of compression algorithm based on statistical similarity is presented that can be effectively used in many applications where an order of data sequence could be relaxed. Next, a quality-adjustable sensor data archiving is discussed, which compresses an entire collection of sensor data efficiently without compromising key features. Considering data aging aspect, this archiving scheme is capable of decreasing data fidelity gradually to secure more storage space.

1 Introduction

Rapid advances of hardware technology have created massive information flow generated by various devices without user activities. For instance, various sensing devices including mobile phones and biomedical sensors are essential nowadays, which generate continuous flows of big sensing data as shown in Fig. 1. In order to handle this, we have to consider how to capture and store sensor data efficiently. As devices gather more and more data, our ability to store data records to persistent storage is significantly challenged. In these cases, compressing the data could provide an effective approach to address a such challenge.

Data compression is a way of representing information in a compact form. This is accomplished by identifying and using structures that exist in the data [51]. A data compression model is categorized as two broad classes: *lossless coding*

D. Lee (✉)
Texas A&M University-Commerce, Commerce, TX 75428, USA
e-mail: dongeun.lee@tamuc.edu

S.C. Suh, T. Anthony (eds.), *Big Data and Visual Analytics*,
https://doi.org/10.1007/978-3-319-63917-8_7

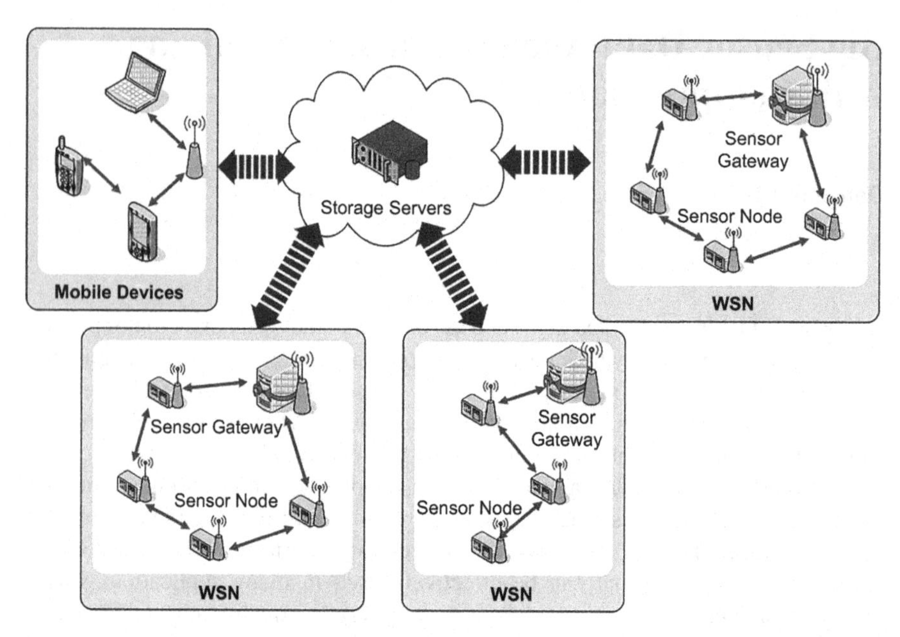

Fig. 1 Data generation scenario from various sensors including mobile phones and sensors in wireless sensor network (WSN)

where a reconstruction of compressed data is identical to the original raw data; and *lossy coding* where a reconstruction is different from the original raw data, while providing much higher compression. Compression schemes discussed in this work mainly benefits from the use of lossy coding.

1.1 Sensor Data Acquisition

An efficient sensing scheme is especially demanding in resource-limited sensors where computational power, network bandwidth, and energy are limited. Apart from the case of resource-limited sensors, full-fledged sensors such as mobile phones can also benefit from the efficient sensing scheme: broader sensing coverage and participation of users are made possible with the efficient sensing. Many conventional distributed sensing schemes [41, 42] process input signals in the sensing devices to reduce the burden of network transmission. However, these conventional schemes are not well suited for resource limited sensing devices because of excessive energy and resource consumptions.

Compressive sensing (CS) [3, 32, 37, 45] sheds light on this problem by shifting the complexity burden of encoding process to the decoder. CS enables to compress large amounts of inputs signals without much energy consumptions. Recent advances in CS reduce the computational burden even further by the random sampling, so that CS schemes are successfully applied to large-scale wireless sensor networks (WSN) [22, 29, 30, 35].

This work proposes a low complexity sampling (LCS) framework that significantly reduces computation and communication overheads of participating sensors [29, 30]. As contrasted with previous CS schemes that demand a certain degree of cooperation among neighboring sensors [3, 37, 45], LCS does not require any additional coordinated action or routing structure on the network of sensors. Since this framework is based on the random sampling, which samples a small part of sensing data and reconstruct them exactly with overwhelming probability when incoming data can be sparsely represented, it substantially reduces the computation and communication overheads involved in sensing operations.

LCS includes a multi-dimensional random sampling when sensor data have sparse representations in spatial and temporal dimensions, which suits the needs of resource-limited sensors and also shows an improved coding efficiency compared to other CS schemes that mostly take account of a single dimension [3, 37, 45].

In general, the effectiveness of a compression method is measured by the compression ratio, which is a ratio between the original data volume and the compressed data volume. To reduce the compressed data volume, we may drop some information in the original data, which is known as lossy compressions [51]. The quality of these techniques is normally judged by the Euclidean distance between the original data and the decompressed version of the compressed data.

However, the focus on Euclidean distance regarding the quality of compression techniques has imposed a significant limitation on the effectiveness of compression methods. In order to break this limitation, we may consider a new type of compression based on a statistical concept known as exchangeability [13], which allows us to capture common data blocks in a more compact form which preserving key properties of data.

Specifically, this approach takes advantage of repeated patterns or common features in the original data, where we consider the distribution of values as the basic pattern. To quantify the similarity of two data blocks, this approach leverages the concept of the locally exchangeable measure (LEM) [13, 36]. When the LEM value passes a given threshold, two blocks are considered to be exchangeable and only one of them is stored as a representative. As more blocks are found to be exchangeable with a stored data block, the existing representative could be kept, which removes the need to store the new blocks. This is a significant departure from common practice in designing compression techniques. This work describes a practical algorithm to realize this unique approach and provides its effective implementation [36].

1.2 Sensor Data Archiving

While data storage capacities keep increasing with reduced cost, faster data generation rate of various devices now leads to a paradox that increasing storage capacity cannot catch up with the rate of information explosion. It is reported that almost half of information created and transmitted cannot be stored now and this mismatch between available storage and information creation will become more serious [25, 27].

From the perspective of an information repository, this mismatch necessitates the development of a big data archiving technique that facilitates scalable and flexible usage of the repository. This work proposes a quality-adjustable archiving scheme for massive sensor data [33, 34]. This scheme thoroughly exploits both spatial and temporal correlations inherent in sensor data collections, and generates a digested set of sensor data keeping fidelity under control, which is demonstrated as outstanding compression efficiency with data fidelity corresponding to orders of sensor accuracies. In addition, a concept of data aging is embodied in the quality-adjustable feature of our scheme with multiple fidelity levels: older sensor data are not as representative as recent data and can be represented with less precision [44].

In distributed environments such as WSN, many approaches have utilized partial correlation to reduce traffic and storage usage inside the networks themselves [7, 23, 24, 38, 49, 58]. Although these approaches have achieved their objectives in distributed environments, efficient archiving techniques are still necessary if sensor data are eventually to be stored in central storage.

A massive amount of data from various sensors should be archived in a cost-effective manner such that the system-wide distortion is minimized under a given storage space. In order to solve this issue, this work proposes analytical models that closely reflect characteristics of our archiving scheme and eventually an optimal storage configuration problem. Since this optimization problem is convex, we can analytically solve it and obtain optimal parameters.

2 Low Complexity Sampling

Most phenomena captured by sensors can be represented with only a few components using approximations of Karhunen-Loève transform such as the discrete cosine transform (DCT) and the wavelet transform [51]. These significant coefficients as well as their position information can be encoded to yield the compact representation of the original signal.

The reason of applying compression schemes on original signal is obvious in the context of dense sensing environment where network bandwidth is a scarce resource: each sensor simply transmitting raw data incurs a bottleneck on the channel toward the collection point (base station in case of wireless sensor networks). When the compression scheme is applied, more sensing data can be transmitted to the collection point.

2.1 Compressive Sensing

However the near-optimal coding of conventional compression schemes is not applicable to many resource constrained devices due to its complexity. Compressive sensing or compressed sampling (CS) can be an option for shifting the complexity burden to the decoder (the collection point, etc.) where original signal is estimated in best-effort manner [4, 11], which can be applied to various types of resource-limited ones such as biosensors.

CS operates very differently from conventional compression schemes as if it were possible to directly acquire just the important information, i.e., significant coefficients of transforms. In CS, a signal is projected onto random vectors whose cardinality is far below the dimension of the signal. For instance, consider a signal $\mathbf{x} \in \mathbb{R}^N$ that can be compactly represented in some orthogonal basis $\mathbf{\Psi}$ with only a few large coefficients and many small coefficients close to zero as follows:

$$\mathbf{x} = \mathbf{\Psi}\mathbf{s}, \tag{1}$$

where $\mathbf{s} \in \mathbb{R}^N$ is the vector of transformed coefficients with a few significant coefficients.

In (1), $\mathbf{\Psi}$ could be any orthogonal basis that makes $\mathbf{x}$ sparse in transform domain such as the DCT and wavelet transform domains. ($\mathbf{\Psi}$ even could be the canonical basis, i.e., the identity matrix $\mathbf{I}$, if $\mathbf{x}$ is sparse itself without the help of transform.) The signal $\mathbf{x}$ is called ***K*-sparse** if it is a linear combination of only $K \ll N$ basis vectors such that $\sum_{i=1}^{K} s_{n_i} \boldsymbol{\psi}_{n_i}$, where $\{n_1, \ldots, n_K\} \subset \{1, \ldots, N\}$; s_{n_i} is a coefficient in $\mathbf{s}$; and $\boldsymbol{\psi}_{n_i}$ is a column of $\mathbf{\Psi}$. Note that a real world signal in general is not exactly K-sparse; rather it can be closely approximated with K basis vectors.

CS projects[1] $\mathbf{x}$ onto a random sensing basis $\mathbf{\Phi} \in \mathbb{R}^{M \times N}$ as follows ($M < N$):

$$\mathbf{y} = \mathbf{\Phi}\mathbf{x} = \mathbf{\Phi}\mathbf{\Psi}\mathbf{s}, \tag{2}$$

where $\mathbf{\Phi}$ is generally constructed by sampling independent identically distributed (i.i.d.) entries from the Gaussian or other sub-Gaussian distributions whose moment-generating function is bounded by that of the Gaussian (e.g., Rademacher distribution).

Consequently, $\mathbf{\Phi}$ is dense with virtually every entry set to non-zero real numbers, which causes two issues: (1) the sensing matrix $\mathbf{\Phi}$ occupies a substantial amount of storage, (2) this leads to $O(MN)$ multiplication and summation operations. Both of these issues can be costly to resource-limited sensors without specific CS-supporting architectures [11].

There has been workarounds for these problems that replace the Gaussian entries of $\mathbf{\Phi}$ with structured random matrices such as the random selection of rows from (forward) Fourier transform matrix [17, 18]. In this case, the signal $\mathbf{x}$ has to be scrambled via a random permutation of its coefficients, in order to bring randomness as in the Gaussian matrix. Using structured random matrices, one need only store indices of randomly selected rows and permutation sequence instead of storing entire entries of Fourier matrix, which is a significant saving for the storage space. In addition, structured random matrices such as Fourier transform and DCT matrices have the fast algorithms available whose computational complexity is $O(N \log N)$ ($\log N < M$ in general).

[1]This can also be seen as inner product operations.

The system shown in (2) is ill-posed as the number of equations M is smaller than the number of variables N: there are infinitely many $\mathbf{x}$'s that satisfy $\mathbf{y} = \mathbf{\Phi x}$. Nevertheless this system can be solved with overwhelming probability provided that $\mathbf{s}$ is sparse and M is large enough such that $M = O(K \log(N/K))$ in the case of Gaussian sensing matrix and $M = O(K \log N)$ in the case of the structured random matrices [22].

2.1.1 General Signal Recovery

A signal recovery algorithm takes measurements $\mathbf{y} \in \mathbb{R}^M$, a random sensing matrix $\mathbf{\Phi}$, and the sparsifying basis $\mathbf{\Psi}$. In a typical setup, the only information an encoder always has to send is $\mathbf{y}$. The sensing matrix $\mathbf{\Phi}$ can be explicitly sent [45] or reconstructed using meta information such as the seed of pseudorandom number generator [19, 37], depending on application. Finally, the sparsifying basis $\mathbf{\Psi}$ is assumed to be known to a decoder. In fact, if a better sparsifying basis is found, then the same $\mathbf{y}$ can be used to reconstruct an even more accurate view of the original signal $\mathbf{x}$.

The signal recovery algorithm then recovers $\mathbf{s}$ knowing that $\mathbf{s}$ is sparse. Once we recover $\mathbf{s}$, the original signal $\mathbf{x}$ can be recovered through (1). It has been shown that the following linear program gives an accurate reconstruction of $\mathbf{s}$:

$$\operatorname{argmin} \|\tilde{\mathbf{s}}\|_1 \qquad \text{subject to} \qquad \mathbf{\Phi \Psi} \tilde{\mathbf{s}} = \mathbf{y}. \tag{3}$$

There are many efficient algorithms that solve (3) using either linear program approaches or iterative, greedy searches [4, 11, 22].

2.1.2 Noisy Signal Recovery

Suppose $\mathbf{y}$ were corrupted with a noise $\mathbf{z} \in \mathbb{R}^M$ that is a stochastic or deterministic unknown error term, which could be from various sources such as communication and quantization. The corrupted $\hat{\mathbf{y}}$ can be represented as

$$\hat{\mathbf{y}} = \mathbf{\Phi \Psi s} + \mathbf{z}. \tag{4}$$

It has been shown that (4) can be solved using the following minimization problem with relaxed constraints for reconstruction:

$$\operatorname{argmin} \|\tilde{\mathbf{s}}\|_1 \qquad \text{subject to} \qquad \|\mathbf{\Phi \Psi}(\mathbf{s} - \tilde{\mathbf{s}}) + \mathbf{z}\|_2 \leq \eta \sqrt{M}, \tag{5}$$

where $\eta \sqrt{M}$ bounds the amount of noise in the signal. Problem (5) is often called LASSO and can also be solved using various efficient algorithms [11, 22].

2.2 *Random Sampling in Spatio-Temporal Dimension*

The sensing/sampling paradigm explained in Sect. 2.1 can be applied to signals that can be represented in one-dimensional vectors, i.e., $\mathbf{x}$. In a distributed sensing context, these vectors not only correspond to time-series data of individual sensors in temporal dimension, but also to data in spatial dimension from a group of sensors at a specific time instant, in which case two-dimensional data can be vectorized into a one dimension.

However, applying the compressive sensing (CS) technique to both spatial and temporal dimensions is not immediately evident. One may want to vectorize one dimension first and the other dimension next, and so on (e.g., vectorize the spatial dimension and then the temporal dimension). Yet this naive approach has problems. First, finding $\mathbf{\Psi}$ that sparsifies $\mathbf{x}$ could be a daunting task. Next, even if we could find a proper order in vectorizing mutually correlated signals, the vectorizing would incur additional complexity that brings control and communication overheads.

Due to these difficulties, most CS literature focused only on one dimension, especially the spatial dimension and proposed their own ways of coordination among distributed sensors to achieve the random measurements of distributed data [3, 37, 45]. (CS in the temporal dimension is straightforward: each sensor can take random measurements in its temporal dimension and send them to the collection point using generic routing schemes.)

Nevertheless, more efficient sensing techniques that exploit the joint correlation of spatio-temporal dimension are solicited for improved coding efficiency. Ignoring another dimension in coding process despite a clear correlation between them results in poor coding efficiency, which can be explained with the help of information theory: the joint entropy rate of two or more random variables is always less than the sum of their individual rates provided that they are dependent on each other [16]. We can achieve the maximal coding efficiency if we leverage correlations and make coding rate (compressed size) close to the joint entropy rate.

As discussed in Sect. 2.1, the random sensing matrix $\mathbf{\Phi}$ in (2) is a dense matrix, which causes complexity in both storage and computation. This can be mitigated with the use of structured random matrices that reduces storage requirement and brings down computational requirement from $O(MN)$ to $O(N \log N)$. Nevertheless, even this requirement could be unacceptable to resource-limited sensors that should handle the rapid generation rate of *big sensing data*.

Thus it is imperative that a more efficient way of sensing is investigated which is far less complex than projecting the original signal onto a *dense* random sensing basis. The sparse random sensing matrix composed of a few ones for each column and zeros for all other entries is one of solutions [8, 32], which has the computational complexity of $O(N \log(N/K))$.

This work utilizes more efficient sensing mechanism based on *random sampling* whose computational time complexity is constant [22]. The random sampling scheme is based on the fact that it is possible to construct $\mathbf{\Phi}$ in (2) from a random selection of rows from the identity matrix $\mathbf{I}$, which is equivalent to the random

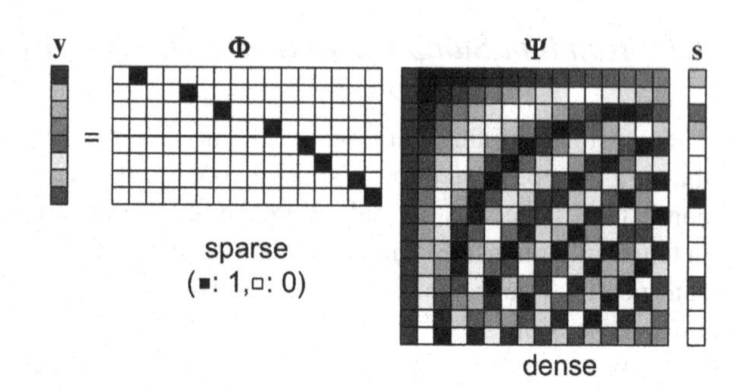

Fig. 2 Random sampling of a signal in (2)

sampling of coefficients in **x**. Note that this scheme works only if the sparsifying basis $\boldsymbol{\Psi}$ is dense such as the DCT and wavelet transform bases, in order not to violate *incoherence*,[2] which is a sufficient condition for the successful recovery of the original signal [11, 22]. The random sampling of signal in the CS setup is illustrated in Fig. 2. Here, the number of required measurements M is the same as in the case of the structured random matrices, that is, $M = O(K \log N)$.

2.2.1 Low Complexity Sampling Framework

Low complexity sampling (LCS) randomly samples the original signal in both spatial and temporal dimensions. First, each sensor randomly samples its time-series data in the temporal dimension using the same indices shared across different sensors, which is equivalent to using the same random sensing matrix $\boldsymbol{\Phi}_{\text{temporal}}$ across different sensors for the same time frame. This sampling reduces the lengths of original time-series data of sensors.

Next, the random sampling is performed in the spatial dimension: the group of randomly sampled coefficients from each sensor is sampled again in the spatial dimension, which is illustrated in Fig. 3. Therefore, in reality each sensor needs to sample and transmit only coefficients that are selected in both spatial and temporal dimensions, as shown in Fig. 3. Here the rationale behind using the same random sampling indices in the temporal dimension is to maximize the spatial correlation of coefficients across different sensors: the spatial correlation is likely to be stronger at the same time instant than with different time instants [30].

LCS is especially useful in the distributed sensing context thanks to the *opt-in* and *opt-out* nature of participating sensors as shown in Fig. 4. In a distributed sensing, each sensor may want to participate in or not depending on its remaining energy [42] or its willingness to volunteer in the context of *participatory sensing* [9]. Although this freedom from structured/coordinated communications can efficiently enable large scale sensing tasks, every sensor acting on the basis of individualism

[2] The two bases $\boldsymbol{\Phi}$ and $\boldsymbol{\Psi}$ are (maximally) incoherent when the largest correlation between any two elements of $\boldsymbol{\Phi}$ and $\boldsymbol{\Psi}$ is $1/\sqrt{N}$ where N is the order of two square matrices.

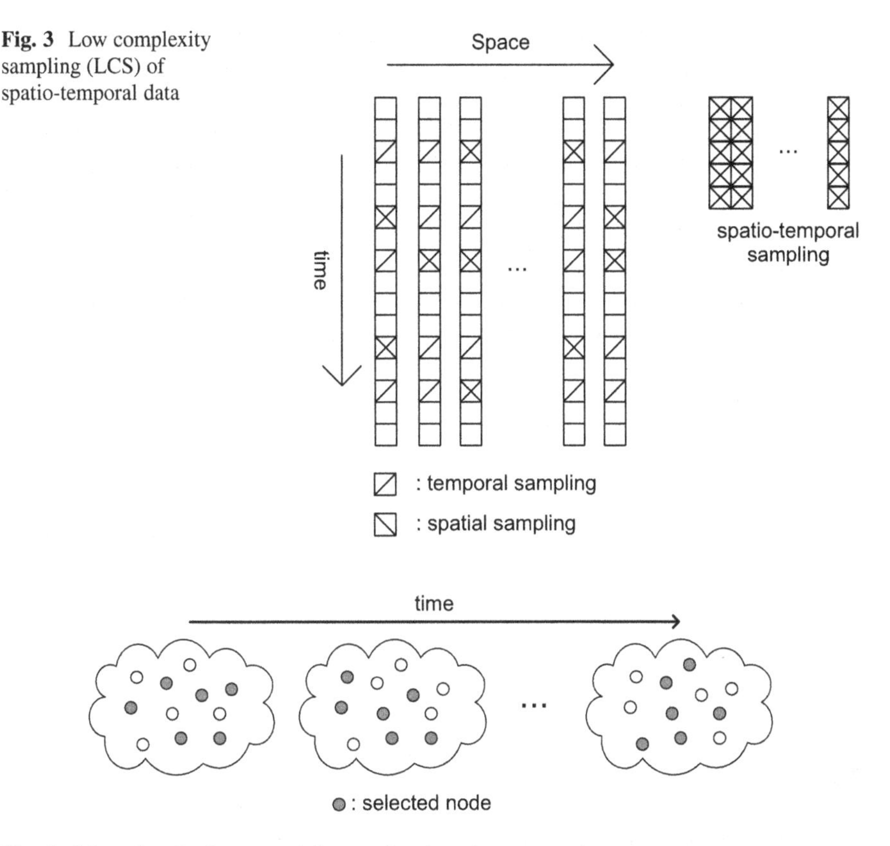

Fig. 3 Low complexity sampling (LCS) of spatio-temporal data

Fig. 4 Selected nodes for transmitting sensing data change over time

may infringe the randomness of sampling in the spatial dimension. Since it is well known that deterministic sensing matrices degrade the performance of compressive sensing [22], the opt-in and opt-out freedom of participating sensors should be carefully exercised.

Regarding how to generate random numbers used for random sampling of spatio-temporal data, popular approaches are to use pseudorandom numbers [19, 37]. These random numbers should be synchronized between encoder (sensors) and decoder (the collection point, etc.), in order to ensure the correct recovery of the original signal as explained in Sect. 2.1.1. Randomness can also be improved via periodically updating random seeds between encoder and decoder. Note that LCS only needs to store random indices of coefficients in spatial and temporal dimensions, instead of storing entries for the random sensing matrix $\mathbf{\Phi}$. This is a significant saving for the storage space of sensors.

Furthermore, the random indices for the spatial sampling need not be explicitly synchronized between encoder and decoder if each sensor determines whether to transmit or not at every time instant that corresponds to the temporal sampling point. The decision can be made based on the opt-in and opt-out policy or the transmission probability defined for each sensor. This is possible because the collection point can recognize the spatial index required for the reconstruction when it receives sampled sensing data from each sensor.

Now we describe the algorithm of LCS. In particular, the operation of each sensor participating in a sensing task can be represented as the following algorithm. (We assume that the random indices for the spatial sampling are explicitly synchronized between encoder and decoder.)

Require: N, temporal random index set $T = \{n_1, \ldots, n_M\} \subset \{1, \ldots, N\}$[3], and spatial random index vector $\mathbf{s} \in \mathbb{R}^M$ ($s_i \in \{0, 1\}$) {$\mathbf{s}$ is the translated version of the spatial random indices from individual sensors' perspective}

1: **loop**
2: $j \leftarrow 1$
3: **for** $i = 1$ to N **do**
4: **if** $i \in T$ **then**
5: **if** $s_j = 1$ **then**
6: transmit sensor reading
7: **end if**
8: $j \leftarrow j + 1$
9: **end if**
10: **end for**
11: **end loop**

2.2.2 Signal Recovery

The collection point of sensing data takes a random sample measurement matrix as described in Fig. 3. Note that the recovery problem in this case is different from conventional CS recovery in the sense that we are dealing with a measurement *matrix*, not a measurement *vector*. Thus we need to decode the measurement matrix.[4]

First, each row of the measurement matrix is decoded following the recovery procedure explained in Sect. 2.1.1. Specifically, the solution $\mathbf{s}^\star$ to (3) obeys

$$\|\mathbf{s}^\star - \mathbf{s}\|_2 \leq C_0 \times \|\mathbf{s} - \mathbf{s}_K\|_1 \tag{6}$$

for some constant C_0, where $\mathbf{s}_K$ is the vector $\mathbf{s}$ with all but the largest K components set to 0: the quality of recovered signal is proportional to that of the K most significant pieces of information. We get progressively better results as we compute

[3]This can be generated by a random permutation.

[4]When the random indices for the spatial sampling are not explicitly synchronized between encoder and decoder, the spatio-temporal measurement no longer has a matrix form since the number of spatial sampling can vary between time instants. However, the decoding process does not impose the matrix form on the spatio-temporal measurement.

more measurements M, since $M = O(K \log N)$ [11]. Therefore, $\mathbf{\Psi}\mathbf{s}^{\star} \in \mathbb{R}^N$ also makes progress on its quality as M increases. (The error bound follows (6) as well if $\mathbf{\Psi}$ is an orthogonal matrix, which is usually the case.)

We now have the half-decoded matrix. Next, each column of the half-decoded matrix is decoded once more to obtain the full-decoded matrix. In contrast to the case of decoding each row of the measurement matrix, decoding each column follows the recovery procedure explained in Sect. 2.1.2, because error occurs during the recovery of spatially sampled coefficients following the error bound in (6).

In particular, the solution $\mathbf{s}^{\star}$ to (5) obeys the following reconstruction error bound:

$$\|\mathbf{s}^{\star} - \mathbf{s}\|_2 \leq C_0 \times \|\mathbf{s} - \mathbf{s}_K\|_1 + C_1 \times \sqrt{K}\eta, \tag{7}$$

where C_1 is another constant for the additional term in the new error bound. Thus, (7) accounts for not only the measurement error due to an insufficient M, but the measurement error carried over from the previous recovery stage, which is explained by η in (5). Accordingly, the total error of our low complexity sensing framework is given by (7).

In order to show the feasibility of LCS, we present the probability of correct signal recovery with varying numbers of measurements and sparsities in Fig. 5, which was obtained by experiments. In particular, we consider an $N \times N$ square matrix whose sparsity is controlled by *artificially* generating signed spikes ± 1 at random locations and then *densifying* each column using $\mathbf{\Psi}$ as in (1). The number of measurements in both spatial and temporal dimensions is controlled by a single parameter M. Note that here we do not deal with real world signals, but synthetic signals. In addition, we consider a perfect signal recovery without error. Therefore the two-stage decoding process only follows the signal recovery procedure explained in Sect. 2.1.1.

In Fig. 5, we can identify that a constant factor for $O(K \log N)$ varies according to the number of measurements: a larger number of measurements translates into a smaller constant factor. In addition, note that there is a transition from success to failure during signal recovery for a given measurement ratio δ. A recovery success analysis with uncertainty in signal sparsity has been recently studied [35].

2.3 Evaluation

LCS has low computation and communication overheads because it is based on random sampling. In particular, the time complexity of the random sampling is constant because it does not involve multiplication and summation operations. The random sampling also uses random indices only for the sampling, in contrast to dense sensing schemes that should store entries for the random sensing matrices.

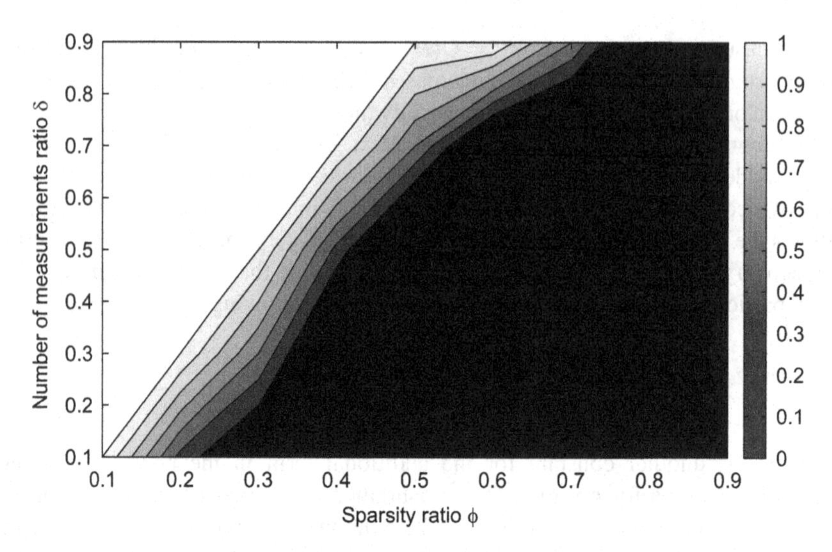

Fig. 5 Probability of correct signal recovery for randomly generated sparse matrices with $N = 256$. The probability is represented as a function of $M = \delta N$ and $K = \phi M$

In addition to the complexity study, one can tell the performance of sensing schemes by comparing the coding efficiency that accounts for how compact data can be compressed with given fidelity. In other words, the sensing scheme with the maximum coding efficiency among others is one that minimizes error between the original signal and the recovered signal with a given data size.

Here we compare LCS with Model-Based CS (block sparse model) [5, 6, 19], which is a state-of-the art CS scheme that takes account of the joint correlation in the spatio-temporal dimension. Model-Based CS is built on the *joint sparsity model* in order to exploit the joint correlation inherent in spatio-temporal dimension. We also include spatial random sampling that performs random sampling in the spatial dimension, as sensing in the spatial dimension is the most popular approach that utilizes CS [3, 37, 45].

Since we are interested in real world sensor data, the experiments employed environmental data sets obtained from wireless sensor network deployment scenarios [46, 54]. In particular, results of two different sensor data types are shown here: (1) ambient temperature (°C) [54] and (2) luminosity (A/W) [46]. These signals are not exactly K-sparse and instead approximated with K basis vectors. Specifically, we utilized DCT throughout the experiments as a sparsifying basis Ψ. As discussed in Sect. 2.2.2, recovered signal quality is improved (i.e., the signal error is decreased) as more random measurements are received by the decoder, because K is increased according to the number of measurements.

Figure 6 shows the experimental results. We here consider sum of squared error (SSE) normalized with respect to the norm of signal as the performance metric. All of the three schemes show decreasing normalized SSE as the number of measurements per individual devices increases. However, how fast normalized

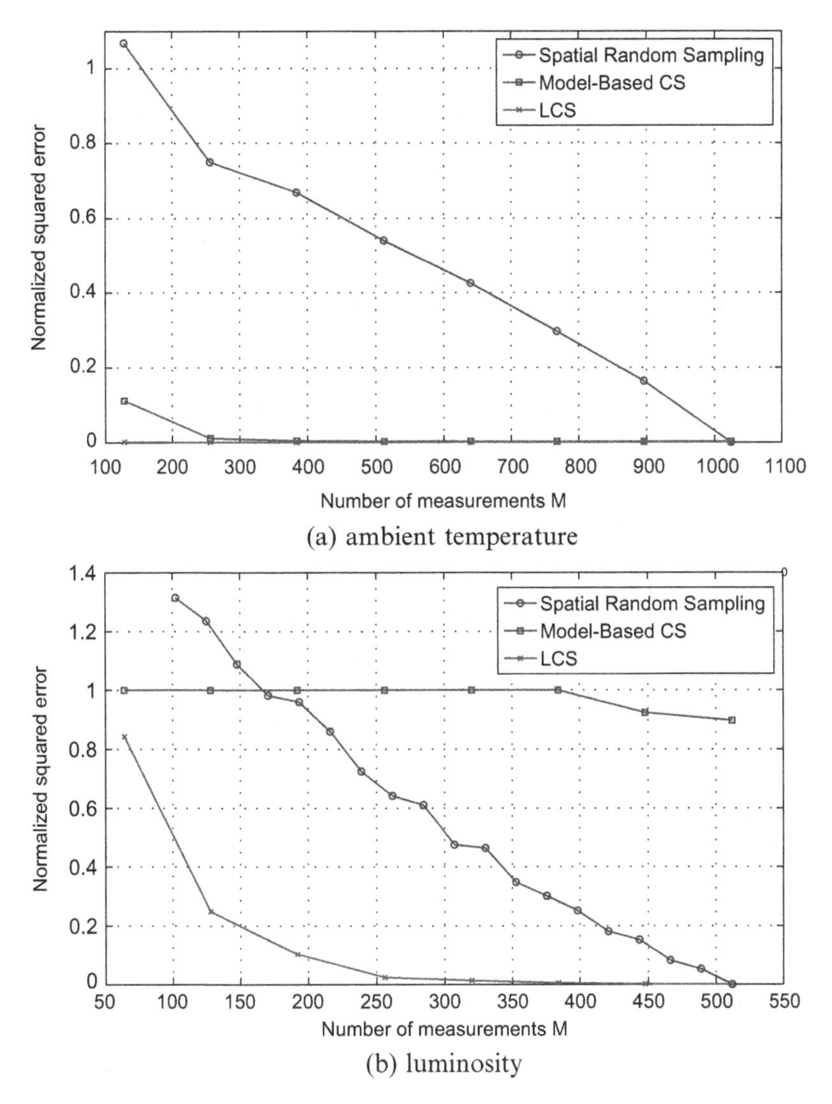

Fig. 6 (**a**) Normalized SSE comparison of ambient temperature data from 32 sensors with the temporal data length $N = 1024$. (**b**) Normalized SSE comparison of luminosity data from 45 sensors with the temporal data length $N = 512$

SSE drops depends on various schemes. We can see that LCS and Model-Based CS both show superior results to the spatial random sampling. Moreover, LCS is superior to Model-Based CS especially in Fig. 6b.

3 Statistical Similarity Based Data Compression

Conventional lossy coding schemes in general quantize or threshold data to adjust quality and reduce data size [51]. Their goal is to compress data without compromising distinctive attributes of data. However, the tenets of these conventional schemes thus far have restricted their attention to the recovery of signal where distortion (distance) is measured using Euclidean distance such as sum of squared error (SSE) and signal-to-noise ratio (SNR) [11, 29, 48, 51]. Specifically, using Euclidean distance as the distance measure requires the sequence of encoded and decoded data to be preserved.

Employing the concept of random variable introduces a new way of signal recovery: data is reconstructed from a learned probability distribution during the encoding process, not from encoded (quality-adjusted) data itself. Thus, encoded output is not a direct representation of original data; instead, the encoder informs the decoder how to regenerate them. If we relax the constraint of preserving the sequence of encoded and decoded data, and treat a sequence of data as if it originates from a random variable, we can achieve a superior compression ratio.

This work presents a new class of compression scheme based on statistical similarity, dubbed IDEALEM (Implementation of Dynamic Extensible Adaptive Locally Exchangeable Measures) [28], that parts with conventional Euclidean distance measure and instead focuses on the exchangeability of similar data sequences [36]. In particular, this flexibility/relaxation on the order of data sequence yields much higher compression ratios.

Of course, application data could not be explained by random numbers. However, in some situations, devices such as sensors might be measuring background noise during the majority of their operation time. In these cases, faithfully reproducing the random noise is not necessary.

3.1 Similarity Measure

Figure 7 shows time-series data of total 64 samples. If we assume that each sequence of 16 samples is an instantiation of a random variable X_i ($i = 1, \ldots, 4$), we can consider similarities between these random variables. In Fig. 7, X_1, X_2, and X_4 all look visually similar; whereas X_3 looks different from other random variables. The design of IDEALEM is based on these observations: we may represent X_1, X_2, and X_4 using a single random variable, claiming that three random variables have an identical distribution behind.

Specifically, if we keep only a single sequence (distribution) from one of three random variables, i.e., X_1, X_2, and X_4, we can achieve data compression with a compression ratio of 3, where the compression ratio is defined by the ratio of an original data size to a compressed data size. Here X_1, X_2, and X_4 are *exchangeable* in the sense that we could represent any of them with each other. Therefore, the more similar random variables we find, the higher compression ratio we can achieve.

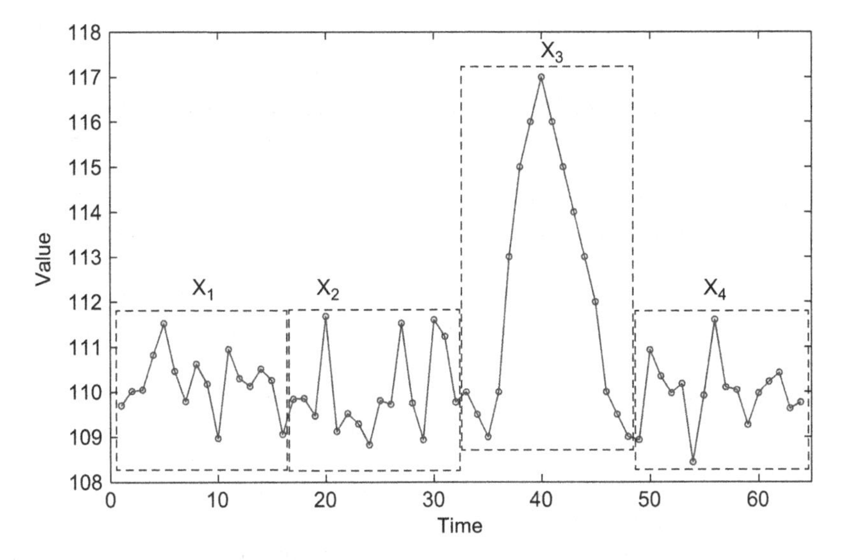

Fig. 7 An example time series and corresponding random variables. All random variables except X_3 were generated from the normal distribution $\mathcal{N}(110, 1)$. Note that X_1, X_2, and X_4 look visually similar, as they share the same underlying probability distribution

Then how can we measure similarity? Based on a given similarity measure, a compression procedure can determine whether X_i is similar to other random variables and then proceed to encode the blocks of data for compression. Kolmogorov-Smirnov test (KS test) is a popular test for the similarity of time series [47, 53, 62]. Thus IDEALEM uses KS test as the similarity measure. Technically speaking, KS test is a non-parametric statistical hypothesis testing method that can test whether two underlying one-dimensional probability distributions of random variables differ or not [20, 39]. In addition to being widely used, KS test is also much easier to compute than related statistical tests such as Anderson-Darling test [20].

To put it simply, as a distributional distance between two random variables grows, the *p-value* yields a smaller value [59]. A small p-value indicates that a *null hypothesis*, i.e., two random variables are from the same distribution, is more likely to be wrong, which automatically supports its logical complement, i.e., two random variables are *not* from the same distribution.

In practice, a threshold α is specified by the user.[5] If a p-value is less than or equal to a chosen α, we reject the null hypothesis, supporting its logical complement. IDEALEM interprets this α as a threshold for similarity, so as to remove redundancy from original data. This does not directly assert whether two random variables are from the same distribution or not; rather, it is a way of identifying similar random variables from the perspective of data compression.

[5]This value is also called the significance level.

3.2 IDEALEM Design

The main idea of IDEALEM is to store only a distribution that is distinct from previous distributions, according to the similarity measure defined with KS test. IDEALEM has three key parameters that affect its operation. First, *block length n* determines the number of samples in an individual sequence. An incoming time series is broken down into blocks with each of them having n elements. Second, *number of buffers b* controls how many distributions are stored in memory for comparison, where each buffer holds one distribution. The number of buffers plays an important role in compression performance: more buffers in general promise higher compression ratios because there is a higher chance of finding a similar distribution stored in buffers when we encounter a new random variable (distribution). However, increasing b also has drawbacks. We cannot simply store too many buffers at the same time in memory, especially if IDEALEM is to be used on resource-limited devices such as sensors. Besides, a new data block is compared against each distribution to compute the p-value. Thus the more buffers we keep, the more KS tests are performed, which increases execution time. Third, *threshold* α explained in Sect. 3.1 is the threshold for similarity when comparing a new random variable to distributions stored in buffers via the KS test. Thus, a lower α results in a higher compression ratio, allowing more random variables to be declared exchangeable. On the other hand, lowering the bar for similarity impairs the reconstruction quality, as it would also include not-so-similar sequences under the same distribution.

3.2.1 Encoded Stream Structure

Figure 8 illustrates an example of encoded stream structure by IDEALEM, where there are three buffers. The first data sequence in an input stream is outputted *as is*, along with the corresponding index which precedes the sequence. Note that this data sequence is also stored in a buffer as the distribution Θ_0. Here, each buffer occupies $8n$ bytes.

Then, the second sequence is encountered and compared against the first sequence. In this example, it is not exchangeable, so the sequence is written on the encoded stream as well as the corresponding index. It is also stored in a buffer as the distribution Θ_1. The third sequence is compared with Θ_0, but not exchangeable. It is next compared with Θ_1, and found to be exchangeable. So the index 1 is solely outputted, where each index takes up 1 byte. The fourth sequence is exchangeable with Θ_1 as well.

The fifth sequence is not exchangeable with any of two stored distributions. So it is again written on the encoded stream as is with the corresponding index. And this sequence also occupies the last remaining buffer as the distribution Θ_2. The sixth sequence is compared with three buffers, but not exchangeable with any of them. Therefore this distribution should be stored in a buffer, which is not immediately

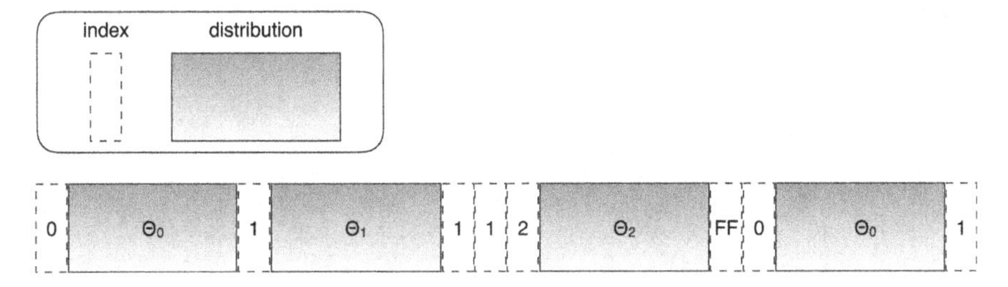

Fig. 8 An encoded stream structure by IDEALEM, where $b = 3$. A *dotted box* represents an index in 1 byte; a *solid box* with gradation a distribution Θ_j. Note that `0xFF` denotes a special marker for overwriting signal

possible since all three buffers are occupied. IDEALEM currently discards the oldest buffer and replaces Θ_0 with this distribution, hence in first-in-first-out (FIFO) manner.

This overwriting should be signaled on the encoded stream so that the decoder can recognize it. To this end, IDEALEM uses a special marker `0xFF`, which automatically limits the number of buffers b to a maximum of 255. This marker is first outputted, and then the index and the sequence is written on the encoded stream. The seventh sequence is compared with from Θ_0 and finally exchangeable with Θ_1. (Comparison with Θ_2 is not necessary.) So only the index 1 is written on the encoded stream.

3.2.2 Decoding

The encoded stream is in turn an input to the decoder of IDEALEM. The decoder reconstructs streaming data from learned probability distributions during the encoding process. This is accomplished with Θ_j's and corresponding indices j's in the encoded stream. For IDEALEM, the relaxation on the order of data sequence entails that it is impossible to reconstruct the same data sequence as the original for exchangeable random variables, i.e., random variables only represented with indices. IDEALEM randomly permutes samples in a learned distribution to generate a new data sequence, which is necessary to avoid any artificial repeated patterns.

3.3 Evaluation

Since the reconstruction quality of IDEALEM cannot be directly assessed using conventional measures such as MSE and SNR, we visually represent reconstruction results with various compression ratios, comparing them with original data. Here it is especially important not to lose *significant patterns* in the original data, which could be abnormal or singular such that they need attention of data analysts. Power grid monitoring data sets from sensors installed on-site at Lawrence Berkeley National Laboratory (LBNL) were used for experiments.

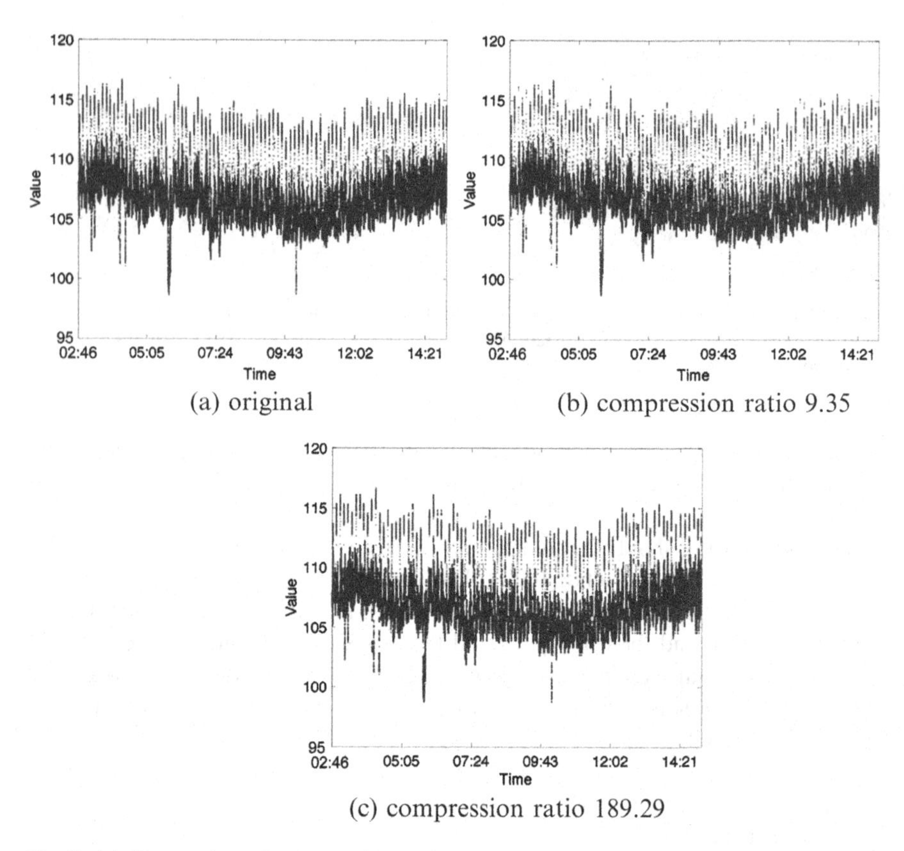

Fig. 9 (**a**) Scatter plots of power grid monitoring data. Reconstructed data from IDEALEM compression with (**b**) $n = 8$, $b = 4$, and $\alpha = 0.1$; (**c**) $n = 32$, $b = 255$, and $\alpha = 0.01$

In Fig. 9, it is obvious that more buffers and lower α mean higher compression ratios. More buffers entail more memory usage and could be a burden to some sensors. However, with $n = 32$ and $b = 255$, the memory usage of IDEALEM is merely 65.28 kB, which is an acceptable amount for resource-limited devices [36]. In Fig. 9, the parameter combination of a compression ratio around 10 and another combination of the maximum compression ratio are shown. We can see the reconstruction quality of IDEALEM is excellent even in the case of the maximum compression ratio 189.29 in Fig. 9c, where all the notable shapes of Fig. 9a are retained. Note that more than 100-fold of compression ratio while capturing important features in data is a property we cannot expect from existing compression schemes. This clearly demonstrates the usefulness of IDEALEM.

4 Scalable Management of Storage Space

In general, individual sensor data do not require either bit-level accuracy or intactness due to several reasons: (1) each sensor node is equipped with inexpensive and imprecise sensors that only guarantee moderate level of sensing accuracy, (2) sensor nodes are densely deployed and they periodically capture environmental data that are highly correlated in spatio-temporal domain, which makes storing all of data unnecessary, (3) we are usually interested in an overall trend of sensor data, thus we can tolerate a certain amount of distortion and approximate results are sufficient most of the time [21, 50, 55]. This property is called the *quality adjustability* in this work.

Data aging, where data fidelity is gradually decreased, is common practice when handling various kinds of time series data [12, 15, 43, 44]. Sensor data fidelity can also be gradually decreased as time goes by. Since fresh data are important (e.g., frequently accessed) and should retain high fidelity, aged data could be regarded less important and only find their use in offering a digest of historical trends in sensor readings. Therefore it is sufficient to store key features of sensor data in most sensor applications especially for long-term storage [23, 24, 57].

Because sensors usually capture physical phenomena such as environmental data, their data are highly correlated in nature within spatial and temporal domain [57]: spatially and temporally close data samples are more correlated than distant counterparts. (Here the degree of correlation is measured by autocorrelation function: one-dimensional in temporal domain and two-dimensional in spatial domain [51].) In particular, the temporal correlation tends to be stronger than the spatial correlation since the sensing frequency of a particular sensor node is in general high enough to surpass the spatial closeness among deployed sensor nodes. This *spatio-temporal correlation*, along with the quality adjustability, allows sensor data to be represented in a compact form.

4.1 Storage Space Optimization

The quality adjustability of sensor data and its trade-off between data fidelity and compression ratio provide us with many options of encoding. Among these numerous operating points, we have to select the best possible way of encoding data that yields the maximum fidelity (the minimum distortion) under a given storage space, i.e., the optimal storage configuration. In other words, we want to solve an optimization problem that requires analytical models, which are unknown. In general, we are not exactly aware of the compressed data size and fidelity prior to an actual encoding that vary depending on a data set. For this reason, we can build new analytical models that are close enough to reflect operating points of archiving scheme, which can be adapted to multiple sensor data types using different model parameters.

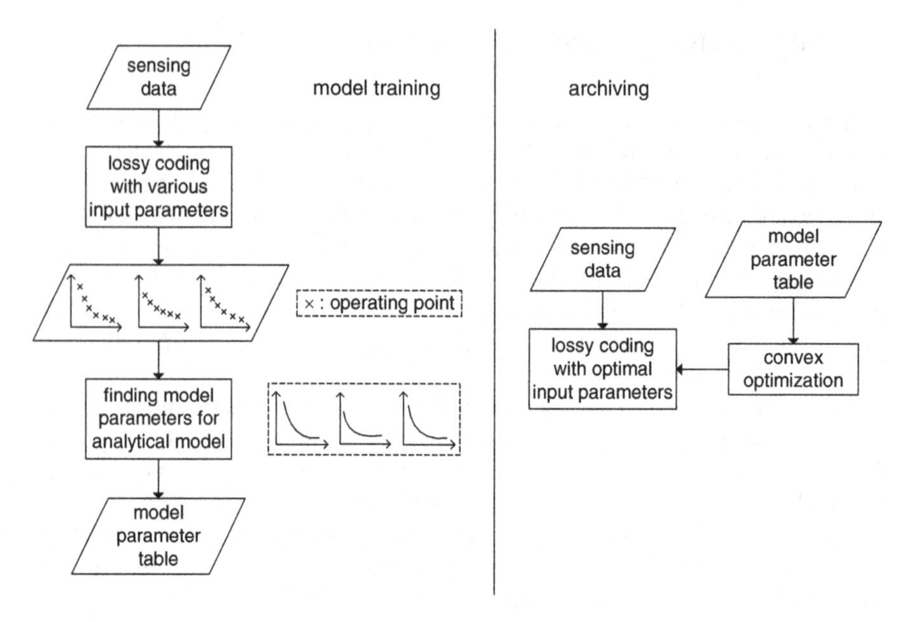

Fig. 10 Flowchart for optimal storage configuration. Each sensor data type has its own model parameter set

Given analytical models, their model parameters are determined when an enough number of data samples for each type of sensor data is gathered. We here assume a stationarity of data for each type without loss of generality, which can be applicable to most sensor data. (For instance, the dynamic range of temperature data does not change over time.) Therefore, we need to determine model parameters only once for each sensor data type in the *model training* process shown in Fig. 10.

Using these model parameters, we can perform a convex optimization that yields the minimum system-wide distortion with a given storage space.[6] The solution of the convex optimization provides optimal input parameters, with which the *archiving* process shown in Fig. 10 is executed. This way, the entire storage space is efficiently utilized.

4.2 Overview of Archiving Scheme

The quality adjustability of sensor data and their spatio-temporal correlation allow to compress the entire data set into a smaller form with a reasonable loss in fidelity. Figure 11 illustrates the block diagram of quality management module,

[6]These analytical models are convex by virtue of the trade-off relationship between data fidelity and compression ratio. The sum of these convex functions is also convex.

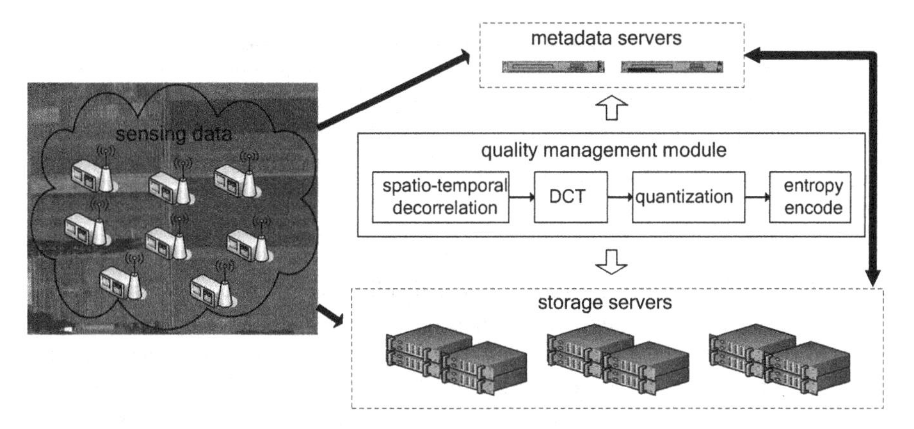

Fig. 11 Quality management module working with conventional distributed file system. *Black arrows* represent data and control flows of file system. Lossy coding runs on storage servers under control of metadata servers

which is designed to work with a conventional distributed file system. Massive data from various sensors are first collected and filtered through the spatio-temporal decorrelation module. Specifically, a sensor value can be predicted by similar values captured by other sensors in close proximity (spatial correlation), or by previous and next similar values captured by that sensor (temporal correlation), whichever is closer than the other, depending on each data instance. If we take a differential between target and predicted values, we ideally obtain a decorrelated value that is close to zero, which means the redundancy in input data is removed.

In reality, these differentiated sensor values still have a fair amount of correlation inside. Therefore the resulting output in turn undergoes the two-dimensional discrete cosine transform (DCT) for further signal decorrelation and energy compaction. DCT is an approximation of Karhunen-Loève transform that is optimal in reducing the dimensionality of feature space [2]. After the two-dimensional DCT, transformed data are subject to the quantization process that sacrifices the precision of data in order to represent them in a compact form, which irrevocably maps a large set of values onto a smaller set. The quantization module controls sensor data fidelity, which can be adjusted through a quantization parameter (QP). The QP determines how much we compress data at the cost of decreased data fidelity. Finally, the entropy encode module compactly produces an encoded data block [51].

4.3 Gradual Decrease of Data Fidelity

In this archiving scheme, multiple temporal levels are supported with a fixed QP. These multiple temporal levels can be utilized as supplementary layers that are gradually discarded as time elapses to incorporate a data aging concept. Figure 12

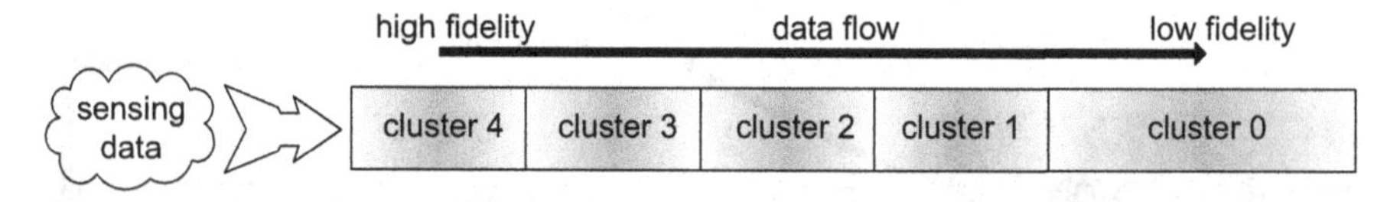

Fig. 12 Quality management module also adjusts temporal quality through the course of sensor data aging

illustrates how incoming sensor data input is handled and archived with the scalable archiving scheme. The quality management module first compresses raw sensor data blocks with a selected QP, which is then stored on the highest fidelity cluster, i.e., the cluster 4 in Fig. 12. When a certain amount of time passes, the quality management module discards the top layer and shift the data block to the next cluster. This process continues until the data block finally reaches the cluster 0, where the data block is archived for a long time. It should be noted that this gradual decrease of data fidelity is analogous to modern scalable video coding schemes [31, 52, 60], which in general enables a smooth transition of video stream from low quality to high quality when there is available bandwidth. On the contrary, the quality management module in this work is designed to gradually degrade data fidelity as data gets old.

4.4 Data Fidelity Model

We can derive analytical models that reflect the effect of adjusting data fidelity on both rate and distortion aspects. This work partially adopts modeling approaches practiced in video coding literature on the one hand [14, 26]. On the other hand, these approaches are limited in modeling every aspect of the archiving scheme, which leads to another unique approach [33, 34]. These new models are shown to be close to actual results, while general models in video coding fail to follow actual results [33]. These analytical models subsequently enable us to develop the optimal storage configuration strategy.

We can model the size of compressed data block R as

$$\begin{aligned} R &= \alpha(\Delta) \times \exp(\beta(\Delta) \times T), \\ \alpha(\Delta) &= a_\alpha \exp(b_\alpha \Delta) + c_\alpha \exp(d_\alpha \Delta), \\ \beta(\Delta) &= a_\beta \exp(b_\beta \Delta) + c_\beta, \end{aligned} \tag{8}$$

where Δ is the quantization step size; $T \in \{0, 1, 2, 3, 4\}$ denotes the temporal level; a_α, b_α, c_α, and d_α are data-dependent constants supplementary to $\alpha(\Delta)$; a_β, b_β, and c_β are constants for $\beta(\Delta)$ [33].

We can also model the amount of distortion D as

$$D(QP, T) = a_{\text{quant}} \exp(b_{\text{quant}} QP) + a_{\text{temp}} T + a_{\text{total}}, \tag{9}$$

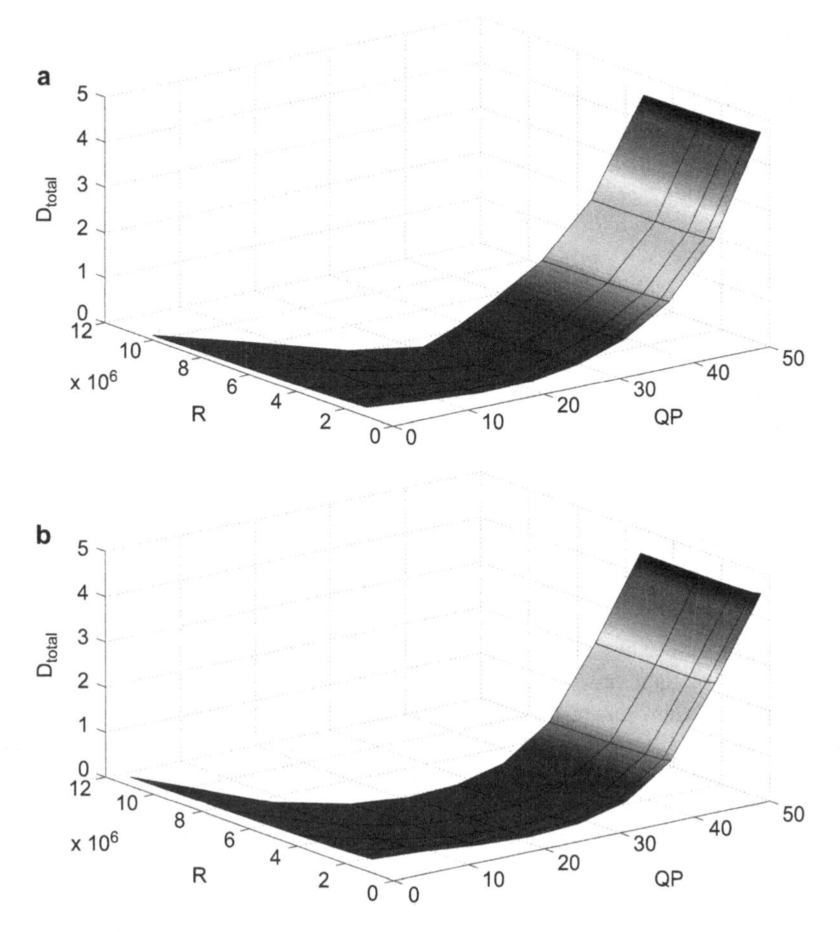

Fig. 13 QP-Rate-Distortion surfaces of ambient temperature data set. Temporal change is implied in the variation of rate. (**a**) Model surface. (**b**) Actual surface

where a_{quant}, b_{quant}, a_{temp}, and a_{total} are data-dependent constants. It should be noted that (9) is a function of QP, whose relationship with the quantization step size Δ is expressed by $\Delta = 0.625 \times 2^{QP/6}$ [14].

The accuracy of the analytical models in (8) and (9) is shown in Fig. 13, where we can identify the analytic models are close to the actual result. Especially in (8) and (9), there is the relationship between QP, temporal level, distortion, and rate, i.e., compressed data size. If we express the relationship without temporal level, we obtain the results shown in Fig. 13a, where the temporal change is implied in the variation of the rate, given a particular QP. The actual QP-Rate-Distortion surface graph is also shown in Fig. 13b for comparison.

4.5 Optimal Rate Allocation

Using the analytical models, our next concern is how to find the minimum distortion with a given specific rate R_0. We first consider an optimal rate allocation problem of single sensor data block, which can be formulated as follows:

$$\begin{aligned} \min_{\{QP,T\}} \quad & D(QP,T) \\ \text{s.t.} \quad & R(QP,T) \leq R_0 \end{aligned}, \tag{10}$$

where $D(QP,T)$ and $R(QP,T)$ is the distortion and the rate function derived in (9) and (8), respectively. We can furthermore extend the rate allocation problem of single sensor data block to accommodate the more general case of storage configuration problem where multiple data blocks have to be stored efficiently [33, 34].

4.6 Evaluation

In order to suggest the efficiency of the scheme, we first compare the compression ratios of popular lossless and lossy coding methods with the scalable data archiving scheme. Sensor data sets are downloaded from the Sensorscope website, which has various wireless sensor network (WSN) deployment scenarios that are mostly environmental data samples [54]. The results convince us that this scheme is a viable solution for archiving a huge amount of sensor data.

Popular lossless coding schemes that were used in experiments for comparison with our scheme are as follows: *gzip*, based on the combination of LZ77 and Huffman coding [56]; *bzip2*, based on the combination of Burrows-Wheeler transform, move-to-front transform, and Huffman coding [10]; *PPMd*, an optimized implementation of prediction by partial matching (PPM) algorithm [40]; Lempel-Ziv-Markov chain algorithm (LZMA), used in *7-Zip* [1]. Figure 14 shows the compression ratios of various schemes that are expressed by the original raw data size divided by the compressed size. Although the compressed size can be as small as how much we allow distortion, it might be unfair to directly compare lossy coding with lossless coding in terms of coding efficiency. Hence we set out a reference point for distortion, which is a sensing accuracy for a particular sensor type that corresponds to the sensor error margin e. Despite an impressive result shown in Fig. 14, total distortion incurred is comparable to the order of sensor error margin e^2.

We also show experimental results with lossy coding schemes that only utilize partial correlation [23, 24, 58]. Wavelet coding is another popular lossy coding method apart from DCT-based coding: it is well known that the performance of wavelet-based and DCT-based codings is almost the same [61]. The results shown in Fig. 15 present the importance of utilizing both spatial and temporal correlations: the wavelet 1D only exploits the temporal correlation for signal compaction, whereas the wavelet 2D only exploits the spatial correlation for signal compaction. After

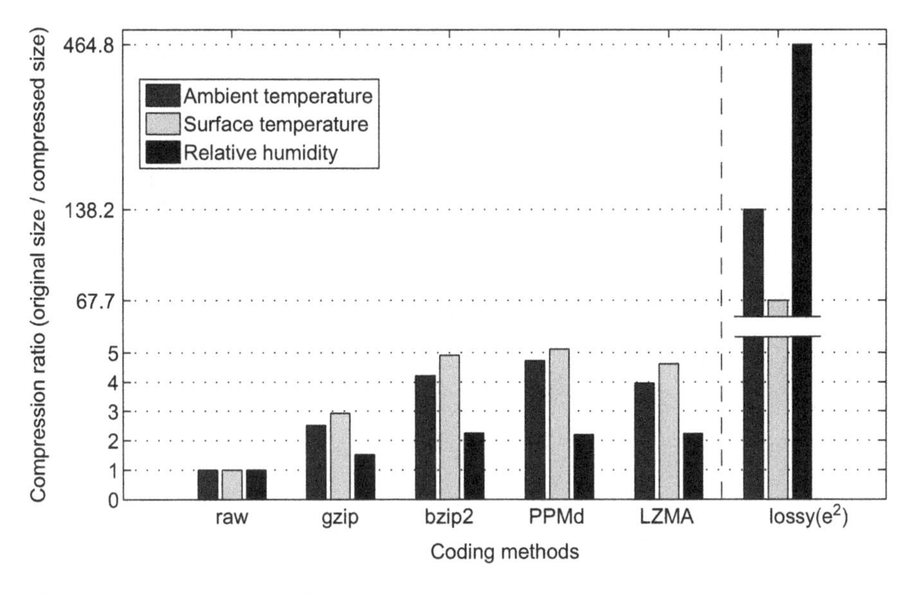

Fig. 14 Compression ratios of the archiving scheme (lossy) compared with various lossless coding methods. We allow distortion up to the sensing accuracies of three different sensor types

signal compaction, both methods apply threshold, quantization and entropy encode processes for the lossy compression of sensor data. Between both wavelet-based methods, the wavelet 1D shows better results than the wavelet 2D, thanks to the stronger correlation in the temporal domain than the spatial domain. Compared to both wavelet methods with partial correlations, our scalable archiving shows much higher compression ratios for three different sensor data sets.

5 Conclusion

This work has reviewed three big data compression techniques that can be utilized for big sensor data acquisition and archiving. We first looked at the low complexity sampling (LCS) framework that can facilitate big sensor data acquisition with low computation and communication overheads. In addition to the low complexity that meets the requirement of resource-limited sensors, LCS shows the improved coding efficiency compared to other compressive sensing schemes. We also discussed a novel data compression technique for data acquisition called IDEALEM. Since IDEALEM is based on statistical similarity, it permits data blocks to be compared without regard to the relative positions of the values in incoming data, leading to higher compression ratios while capturing important features in data.

Next, a quality-adjustable sensor data archiving was discussed, which compresses an entire collection of sensor data efficiently without compromising key

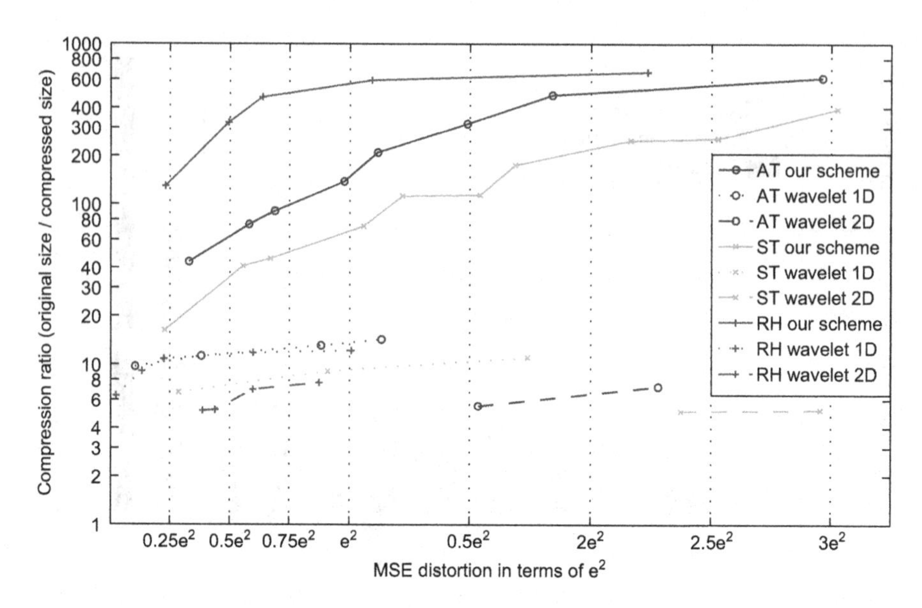

Fig. 15 Log-scale compression ratios of our archiving scheme compared with wavelet-based methods using partial correlations at various data fidelities in terms of the sensing accuracy. Results of ambient temperature (AT), surface temperature (ST), and relative humidity (RH) are shown

features. Considering the data aging aspect, this archiving scheme is capable of decreasing data fidelity gradually to secure more storage space. In order to analyze the performance of the archiving scheme, analytical models that reflect the effect of adjusting data fidelity on both rate and distortion aspects were introduced. And these models helped to obtain a rate allocation strategy that can minimize the system-wide distortion under a given storage space.

References

1. 7-zip. http://www.7-zip.org
2. Ahmed, N., Natarajan, T., Rao, K.R.: Discrete cosine transform. IEEE Trans. Comput. **C-23**(1), 90–93 (1974)
3. Bajwa, W., Haupt, J., Sayeed, A., Nowak, R.: Compressive wireless sensing. In: Proceedings of the International Conference on Information Processing in Sensor Networks (IPSN '06), pp. 134–142 (2006)
4. Baraniuk, R.G.: Compressive sensing [lecture notes]. IEEE Signal Process. Mag. **24**(4), 118–121 (2007)
5. Baraniuk, R.G., Cevher, V., Duarte, M.F., Hegde, C.: Model-based compressive sensing. IEEE Trans. Inf. Theory **56**(4), 1982–2001 (2010)
6. Baron, D., Duarte, M.F., Wakin, M.B., Sarvotham, S., Baraniuk, R.G.: Distributed compressive sensing (2009). arXiv preprint arXiv:0901.3403
7. Barr, K.C., Asanović, K.: Energy-aware lossless data compression. ACM Trans. Comput. Syst. **24**(3), 250–291 (2006)

8. Berinde, R., Indyk, P.: Sparse recovery using sparse random matrices. Tech. Rep. MIT-CSAIL-TR-2008–001, Massachusetts Institute of Technology (2008)
9. Burke, J.A., Estrin, D., Hansen, M., Parker, A., Ramanathan, N., Reddy, S., Srivastava, M.B.: Participatory sensing. Center for Embedded Network Sensing (2006)
10. bzip2. http://www.bzip.org
11. Candès, E.J., Wakin, M.B.: An introduction to compressive advanced sampling. IEEE Signal Process. Mag. **25**(2), 21–30 (2008)
12. Cao, H., Wolfson, O., Trajcevski, G.: Spatio-temporal data reduction with deterministic error bounds. VLDB J. **15**(3), 211–228 (2006)
13. Choi, J., Hu, K., Sim, A.: Relational dynamic Bayesian networks with locally exchangeable measures. Tech. Rep. LBNL-6341E, Lawrence Berkeley National Laboratory (2013)
14. Coding of audiovisual objects - Part 10: advanced video coding (2003)
15. Cohen, E., Kaplan, H.: Aging through cascaded caches: performance issues in the distribution of web content. In: Proceedings of the 2001 ACM Conference on Special Interest Group on Data Communication, pp. 41–53 (2001)
16. Cover, T.M., Thomas, J.A.: Elements of Information Theory, 2nd edn. Wiley, Hoboken (2006)
17. Do, T.T., Gan, L., Nguyen, N.H., Tran, T.D.: Fast and efficient compressive sensing using structurally random matrices. IEEE Trans. Signal Process. **60**(1), 139–154 (2012)
18. Duarte, M.F., Wakin, M.B., Baraniuk, R.G.: Fast reconstruction of piecewise smooth signals from incoherent projections. In: Proceedings of the Workshop Signal Processing with Adaptive Sparse Structured Representations (SPARS '05) (2005)
19. Duarte, M.F., Wakin, M.B., Baron, D., Baraniuk, R.G.: Universal distributed sensing via random projections. In: Proceedings of the International Conference on Information Processing in Sensor Networks (IPSN '06), pp. 177–185 (2006)
20. Engmann, S., Cousineau, D.: Comparing distributions: the two-sample Anderson-Darling test as an alternative to the Kolmogorov-Smirnoff test. J. Appl. Quant. Methods **6**(3), 1–17 (2011)
21. Esmaeilzadeh, H., Sampson, A., Ceze, L., Burger, D.: Architecture support for disciplined approximate programming. In: Proceedings of the International Conference on Architectural Support for Programming Languages and Operating Systems (ASPLOS '12), pp. 301–312 (2012)
22. Foucart, S., Rauhut, H.: A Mathematical Introduction to Compressive Sensing. Springer, New York (2013)
23. Ganesan, D., Estrin, D., Heidemann, J.: Dimensions: why do we need a new data handling architecture for sensor networks? SIGCOMM Comput. Commun. Rev. **33**(1), 143–148 (2003)
24. Ganesan, D., Greenstein, B., Estrin, D., Heidemann, J., Govindan, R.: Multiresolution storage and search in sensor networks. Trans. Storage **1**(3), 277–315 (2005)
25. Gantz, J.F., Chute, C., Manfrediz, A., Minton, S., Reinsel, D., Schlichting, W., Toncheva, A.: The diverse and exploding digital universe: an updated forecast of worldwide information growth through 2011. White Paper (2008)
26. Hang, H.M., Chen, J.J.: Source model for transform video coder and its application. I. Fundamental theory. IEEE Trans. Circuits Syst. Video Technol. **7**(2), 287–298 (1997)
27. Hilbert, M., López, P.: The world's technological capacity to store, communicate, and compute information. Science **332**(6025), 60–65 (2011)
28. IDEALEM. http://datagrid.lbl.gov/idealem
29. Lee, D., Choi, J.: Low complexity sensing for big spatio-temporal data. In: Proceedings of the International Conference on Big Data (BigData '14), pp. 323–328 (2014)
30. Lee, D., Choi, J.: Learning compressive sensing models for big spatio-temporal data. In: Proceedings of the International Conference on Data Mining (SDM '15), pp. 667–675 (2015)
31. Lee, D., Lee, Y., Lee, H., Lee, J., Shin, H.: Determining efficient bit stream extraction paths in H.264/AVC scalable video coding. In: Proceedings of the Asilomar Conference on Signals, on Systems, and Computers (ACSSC '08), pp. 2233–2237 (2008)
32. Lee, D., Choi, J., Shin, H.: Low-complexity compressive sensing with downsampling. IEICE Electron. Express **11**(3), 20130947 (2014)

33. Lee, D., Choi, J., Shin, H.: A scalable and flexible repository for big sensor data. IEEE Sensors J. **15**(12), 7284–7294 (2015)
34. Lee, D., Ryu, J., Shin, H.: Scalable management of storage for massive quality-adjustable sensor data. Computing **97**(8), 769–793 (2015)
35. Lee, D., Lima, R., Choi, J.: Improving imprecise compressive sensing models. In: Proceedings of the Thirty-Second Conference on Uncertainty in Artificial Intelligence (UAI '16), pp. 397–406 (2016)
36. Lee, D., Sim, A., Choi, J., Wu, K.: Novel data reduction based on statistical similarity. In: Proceedings of the International Conference on Scientific and Statistical Database Management (SSDBM '16), pp. 21:1–21:12 (2016)
37. Luo, C.,Wu, F., Sun, J., Chen, C.W.: Compressive data gathering for large-scale wireless sensor networks. In: Proceedings of the Mobile Computing and Networking (MobiCom '09), pp. 145–156 (2009)
38. Marcelloni, F., Vecchio, M.: Enabling energy-efficient and lossy-aware data compression in wireless sensor networks by multi-objective evolutionary optimization. Inf. Sci. **180**(10), 1924–1941 (2010)
39. Massey, F.J. Jr.: The Kolmogorov-Smirnov test for goodness of fit. J. Am. Stat. Assoc. **46**(253), 68–78 (1951)
40. Moffat, A.: Implementing the PPM data compression scheme. IEEE Trans. Commun. **38**(11), 1917–1921 (1990)
41. Noh, D., Lee, D., Shin, H.: Mission-oriented selective routing for wireless sensor networks. In: Proceedings of the International Conference on Communications and Networking in China (CHINACOM '07), pp. 809–813 (2007)
42. Noh, D., Lee, D., Shin, H.: QoS-aware geographic routing for solar-powered wireless sensor networks. IEICE Trans. Commun. **90**(12), 3373–3382 (2007)
43. Palmer, M.: Seven principles of effective RFID data management (2005)
44. Palpanas, T., Vlachos, M., Keogh, E., Gunopulos, D., Truppel, W.: Online amnesic approximation of streaming time series. In: Proceedings of the International Conference on Data Engineering (ICDE '04), pp. 339–349 (2004)
45. Quer, G., Masiero, R., Munaretto, D., Rossi, M., Widmer, J., Zorzi, M.: On the interplay between routing and signal representation for compressive sensing in wireless sensor networks. In: Proceedings of the Information Theory and Applications Workshop (ITA '09), pp. 206–215 (2009)
46. Quer, G., Masiero, R., Pillonetto, G., Rossi, M., Zorzi, M.: Sensing, compression, and recovery for WSNs: sparse signal modeling and monitoring framework. IEEE Trans. Wireless Commun. **11**(10), 3447–3461 (2012)
47. Quinsac, C., Basarab, A., Girault, J.M., Kouamé, D.: Compressed sensing of ultrasound images: sampling of spatial and frequency domains. In: Proceedings of the International Workshop on Signal Processing Systems (SiPS '10), pp. 231–236 (2010)
48. Richardson, I.E.: The H.264 Advanced Video Compression Standard, 2nd edn. Wiley, Hoboken (2010)
49. Sadler, C.M., Martonosi, M.: Data compression algorithms for energy-constrained devices in delay tolerant networks. In: Proceedings of the International Conference on Embedded Network Sensor Systems (SenSys '06), pp. 265–278 (2006)
50. Sampson, A., Nelson, J., Strauss, K., Ceze, L.: Approximate storage in solid-state memories. In: Proceedings of the International Symposium on Microarchitecture (MICRO '46), pp. 25–36 (2013)
51. Sayood, K.: Introduction to Data Compression, 4th edn. Morgan Kaufmann, Burlington (2012)
52. Schwarz, H., Marpe, D., Wiegand, T.: Overview of the scalable video coding extension of the H.264/AVC standard. IEEE Trans. Circuits Syst. Video Technol. **17**(9), 1103–1120 (2007)
53. Seabra, J., Sanches, J.: Modeling log-compressed ultrasound images for radio frequency signal recovery. In: Proceedings of the International Conference on Engineering in Medicine and Biology Society (EMBC '08), pp. 426–429 (2008)

54. Sensorscope: Sensor networks for environmental monitoring. http://lcav.epfl.ch/op/edit/sensorscope-en
55. Srisooksai, T., Keamarungsi, K., Lamsrichan, P., Araki, K.: Practical data compression in wireless sensor networks: a survey. J. Netw. Comput. Appl. **35**(1), 37–59 (2012)
56. The gzip home page. http://www.gzip.org
57. Vuran, M.C., Akan, Ö.B., Akyildiz, I.F.: Spatio-temporal correlation: theory and applications for wireless sensor networks. Comput. Netw. **45**(3), 245–259 (2004)
58. Wang, Y.C., Hsieh, Y.Y., Tseng, Y.C.: Multiresolution spatial and temporal coding in a wireless sensor network for long-term monitoring applications. IEEE Trans. Comput. **58**(6), 827–838 (2009)
59. Wasserstein, R.L., Lazar, N.A.: The ASA's statement on p-values: context, process, and purpose. Am. Stat. **70**(2), 129–133 (2016). doi:10.1080/00031305.2016.1154108
60. Wien, M., Cazoulat, R., Graffunder, A., Hutter, A., Amon, P.: Real-time system for adaptive video streaming based on SVC. IEEE Trans. Circuits Syst. Video Technol. **17**(9), 1227–1237 (2007)
61. Xiong, Z., Ramchandran, K., Orchard, M.T., Zhang, Y.Q.: A comparative study of DCT-and wavelet-based image coding. IEEE Trans. Circuits Syst. Video Technol. **9**(5), 692–695 (1999)
62. Yu, J., Ongarello, S., Fiedler, R., Chen, X., Toffolo, G., Cobelli, C., Trajanoski, Z.: Ovarian cancer identification based on dimensionality reduction for high-throughput mass spectrometry data. Bioinformatics **21**(10), 2200–2209 (2005)

Advanced High Performance Computing for Big Data Local Visual Meaning

Ozgur Aksu

Abstract Being able to scale interactive analysis for big data clusters is gaining more importance with each passing day in our present time. For example, according to the Behaviors Questionnaire performed in 2015 by K.D. Nuggets, around one fourth of 459 participants tried to interpret data clusters that exceed 1 terabyte and 100 petabytes. One of the subjects of previous studies is the Canonic Method, which is used to form the meanings of big data in a fast and efficient manner, because approximate responses given as based on sampling usually bring benefits as much as the response itself; and the sampling may also lessen the burden of cognitive confusion in meaning. As a result of previous studies conducted on database environments, extremely precious data have been obtained in terms of sampling and local visual inference; however, in the present study, firstly, the new methods and the system problems on the access to inference data have been focused on [1–4].

Today, data production is developing at an amazing speed. In our present day, the exponential technical developments, analogue sensor data, adaptive digital systems, scientific high-sensitivity sensors, smart devices and integral-theoretical models cause that data are produced at an extremely great speed. It is expected that global data volume will grow at a speed of 40-fold each year and reach 44 zettabytes by 2020 [5]. The term "big data" has been produced in order to cope with the volume, speed and variety of the data produced, and to make sense of this data trend that is developing day by day. Big data are becoming the new focal point of technology in many fields. A series of additional tools and mechanisms may be integrated to big data systems in order to obtain, store and process different data. These systems use the advantage of a tremendous parallel processing power for the purpose of performing complex conversions and analyses. On the other hand, designing and using a big data system intended for a certain application is not practical [6–7], because data come from more than one source that are heterogeneous and autonomous, and are in complex and changing relations with each other growing in an adaptive manner. In addition to these, the rise of big data applications in

O. Aksu (✉)
University of Alabama at Birmingham, Birmingham, AL, USA

Erciyes University, Kayseri, Turkey
e-mail: oaksu@uab.edu; oaksu@erciyes.edu.tr

S.C. Suh, T. Anthony (eds.), *Big Data and Visual Analytics*,
https://doi.org/10.1007/978-3-319-63917-8_8

which data collection phenomenon is increasing at an amazing speed is beyond the capacity of today's hardware and software platforms in terms of managing, storing and processing data within a reasonable time [6].

1 Introduction

Being able to scale interactive analysis for big data clusters is gaining more importance with each passing day in our present time. For example, according to the Behaviors Questionnaire performed in 2015 by K.D. Nuggets, around one fourth of 459 participants tried to interpret data clusters that exceed 1 terabyte and 100 petabytes. One of the subjects of previous studies is the Canonic Method, which is used to form the meanings of big data in a fast and efficient manner, because approximate responses given as based on sampling usually bring benefits as much as the response itself; and the sampling may also lessen the burden of cognitive confusion in meaning. As a result of previous studies conducted on database environments, extremely precious data have been obtained in terms of sampling and local visual inference; however, in the present study, firstly, the new methods and the system problems on the access to inference data have been focused on [1–4].

Today, data production is developing at an amazing speed. In our present day, the exponential technical developments, analogue sensor data, adaptive digital systems, scientific high-sensitivity sensors, smart devices and integral-theoretical models cause that data are produced at an extremely great speed. It is expected that global data volume will grow at a speed of 40-fold each year and reach 44 zettabytes by 2020 [5]. The term "big data" has been produced in order to cope with the volume, speed and variety of the data produced, and to make sense of this data trend that is developing day by day. Big data are becoming the new focal point of technology in many fields. A series of additional tools and mechanisms may be integrated to big data systems in order to obtain, store and process different data. These systems use the advantage of a tremendous parallel processing power for the purpose of performing complex conversions and analyses. On the other hand, designing and using a big data system intended for a certain application is not practical [6–7], because data come from more than one source that are heterogeneous and autonomous, and are in complex and changing relations with each other growing in an adaptive manner. In addition to these, the rise of big data applications in which data collection phenomenon is increasing at an amazing speed is beyond the capacity of today's hardware and software platforms in terms of managing, storing and processing data within a reasonable time [6] (Fig. 1).

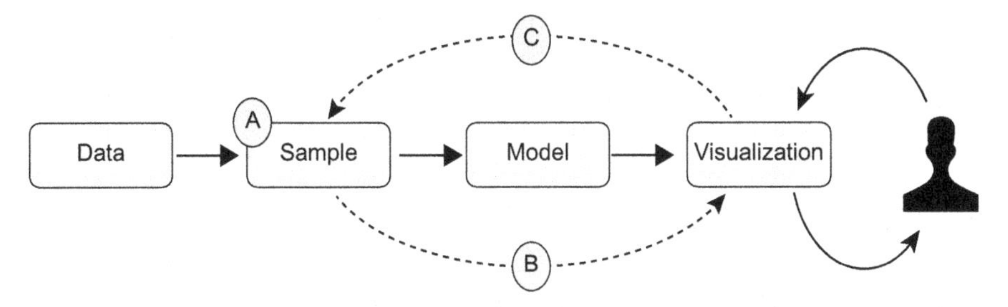

Fig. 1 Visual meaning framework with rapid sampling. (**a**) sampling methods, (**b**) sample visualizations, and (**c**) interactivity between the user and the sampling process [4]

2 The Impact of Latency in Rapid Prototype Applications

According to Liu and Heer [8], a tiny delay of 500 ms may create significant effect on the inference analysis. It is in such a great extent that such delays affect how much of online data clusters are discovered and the final collection speed; and negative externalities about the delays also continue after the removal of the delay. On the very basis of the present study, there is the behavior of the experience of great amounts of data to the reduction of data as well as the realization of the experiences in a slow manner, which is a common problem observed in adaptive speed applications. Although data size grows at such an enormous speed, the response sensitivity is damaged. Query times increase and the Adaptive Simulation Reaction Time delays the response. In such situations, the intelligence software at the front end holding the database at the rear end and vice-versa is a result inference that is frequently observed. Both solutions holding the network responsible for this is more frequently observed. The majority of inference and visualization platforms were designed for gigabyte technology in late 2010 and for early 2020. For this reason, these platforms try to perform their duties in the terabyte and petabyte economy used commonly today. What is more important is that solution-focused institutions do not welcome delay-sourced problems. Delay causes that a series of bad behaviors emerges becoming more serious within the products with the emergence of analogue data scientists in simulation behaviors in time. Similar problems are observed in the development processes of electronic systems, and the problematic data field is growing with each passing day in terms of the design and solution expectations. Developing a solution in possible solution alternatives and solution space developing problem tree is becoming more and more difficult, and there appear losses in the value of data in terms of the relevant processes. Our research has uncovered four impacts of latency for Analogue Rapid Systems:

1. Making the system become parallel for a higher system success (High Calculation Effort),
2. Using simulation software in the design process (Pre-emulation),

3. Using an artificial intelligence algorithm that is adapted for the purpose (Adaptive Artificial Intelligence), and
4. Online assessment of the design result (Online Simulation for total span).

3 Big Data on Analog Electronic Design Challenge

Circuits and signal processing research looks to enhance the speed, reliability, energy-efficiency, manufacturability, and computer-aided design of circuits and systems. Design, modeling, and testing are integral parts of this process. Current research projects include robust, low-power nanoscale integrated circuit design, systems-on-a-chip design for wireless communications, DSP-enhanced high-speed mixed-signal integrated circuit design, and the physics of integrated circuit failure. Work on computer-aided design includes hardware/software co-design, high-level and logic synthesis, compact modeling, physical design, and design for manufacturing. Technology interacts with the world by detecting, transcoding, understanding and generating time-dependent and space-dependent signals in communication, sensor processing, low voltage signal process, and power consumption. Analog circuit processing is a discipline of applied mathematics, concerned with the optimal detection, transcoding, understanding, and generation of signals. Current research projects include high-accuracy and/or high-efficiency processing of speech, audio, image and video signals, computer intelligent interaction, and visual meaning in domains including new concept controllers, sensors, free-field CPU, bioelectric, and biomedical systems. The design of AE (Analog Electronic) circuits is a complicated and hard design problem [9–12]. AE design approaches are not considered as having reached adequate level when this problem, which is defined as the Analog Dilemma, is compared with the digital solution approaches [10, 13]. In order to receive or produce any physical signal, it is absolutely necessary to have electronic circuits, which clearly indicates that there will always be a demand for AE circuits at basic level for such circuits [9, 13–15]. The EHW automations developed by the researchers gain importance with the solution approach in Analog Circuit design works [10, 16]. EHW ensures that electronic circuits that are beyond the reach of manual design works of humans are designed by using evolutionary algorithms [15, 17]. The EHW Architecture brings Evolvable Hardware, Artificial Intelligence, error tolerance and automatic design systems together [11]. The Law of Moore defines that the intra-integrated circuit capacity will double in every 18 months [18]. In our present day, this law continues to confirm itself, and complicates the design processes with exponential increase. The EHW Fast Prototyping approach provides an important acceleration in traditional AE Design Process with FPAA Solution Model [19]. This acceleration provides indirect solutions in problems to which humans cannot provide direct solutions with their skills and capacities with fast prototyping models. The use of the proposed solution methodology in the design process is given in Fig. 2 in the scope of the problem detected.

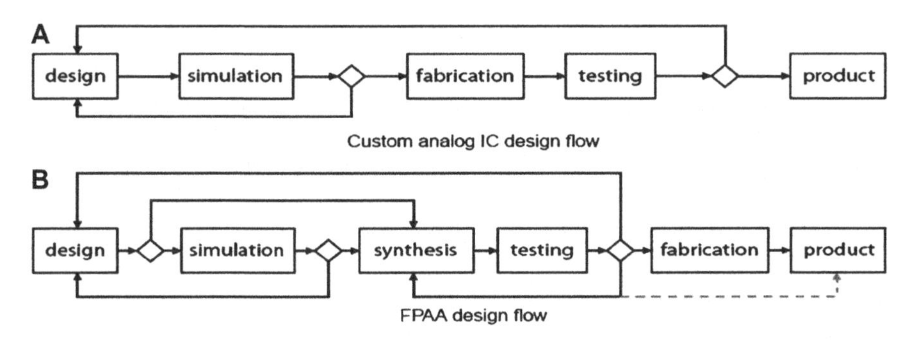

Fig. 2 Custom analog IC design (**a**) FPAA design flow (**b**) [19]

In this study, firstly, the AE design process problem is defined. The studies conducted on ADA Fast Prototyping for the defined problem were examined in the literature for the years 1980–2016, and are given in a comparative manner in Table 1. These studies are examined in terms of the proposed solution methods and sample applications on the defined problem in the literature, and the missing points in the process and the problems are highlighted. In this study, as the final item, a new solution model and a sample application are given based on the motivation areas obtained. There are many similar solution offers on the proposed topic and on the solution to the problems detected in similar solution offers in the scope of the GENAN Model. It is targeted that the innovative potential of the developed solution approach on EHW AE automation is revealed again for today's researchers [14, 20].

ACAD/CAD software and tools have been used actively as of early 1970s [10, 21]. The solution models used provide solutions that are specific to the target by bringing the proper circuit models and functions together [17, 22]. In case there are no models or functions for the solution intended to be developed, it is necessary that firstly the subordinated circuit models are developed for the targeted aim [9, 10, 18, 23, 24]. The subordinated circuit used today has been produced based on the requirements in the electronic libraries [10]. Active and passive AE circuit elements are used in order to form a circuit model that will give the targeted aim in the development process [9, 10, 15, 17, 22]. However, the issue of which of the electronic circuit elements that are plenty in number and type will be used for the targeted aim is a basic research problem [9, 10]. In addition to the AE Element Selection problem; The question of with which connection structure the circuit elements would be connected to each other and in which order constitutes an extremely difficult engineering problem [9, 10, 18, 23, 25]. In order to resolve this problem, scientists separate the frame of electronic science design process into two parts; AE and DE. The basic reason of this separation is developing a new electronic language and library in which it is possible to conduct design/application

Table 1 Analog circuit design challenges [20, 37]

Topic	Digital electronic	Analog electronic	Comment
Circuit design	With the binary system and logic equations	Fully in the actual physical quantities	Analog circuit design is difficult
Circuit elements	Logic gates and libraries (10^3 pieces)	Passive and active discrete components (∞ units)	Number of analog circuit element is not known even now
Circuit size	No any factors within an integrated approach	This is a factor because circuit cannot beget integrated	Analog circuit must be taken into integrated
Placement	Automatic placement and routing tools are available	Placement and routing criteria include multi-parameter	Analog circuits have more difficult layout problem
VLSI design	Is based on a very practical and easy for the library	For there is no any library that covers all elements, solution is specific	Analog circuits need to standardize for VLSI design
Power consumption	Logic components contains many analog sub-elements, therefore power consumption is high and the complex structure	For used less and simple circuit element, power requirement is lower	Analog circuits have low power consumption
Working conditions	For high speed requirement, working conditions is specific	Operating conditions terms is capable of independent operation	Working conditions of analog circuits are more flexible than digital circuits
Practical design tools	FPGA	FPMA, FPTA, FPAA	Integrated design tools for analog circuits has been developed within the dynamically programmed
Production costs	Library-based approach for the production large number of sub-circuit unit requires	Less cost-effective because they have less contain circuit elements and less circuit size	Much more suited to the production costs of analog circuits

with basic rules and which has less complicated structure for the purpose of defining and computing the problems of AE [9, 14, 26]. The DE that is developed over the rules is fictionalized completely in a structure that works on binary number system [9, 14, 26]. Although DE works on AE rules and elements, its design and computing processes are in a simpler structure [14, 26, 27]. The EHW viewpoint has been developed in electronic science as a result of the development of DE-targeted automatic design tools and as a result of the artificial intelligence support [10, 16, 27]. EHW solution processes and models ensure that humans develop practical solutions and perform automatic design processes in the complicated design process [10, 16, 27]. The solution proposal that is obtained with EHW Models require that the solution is renewed as the targeted aim details increase and as more demands for details develop over the result [10, 27, 28]. The developing details of the solution offer increase the number of the elements of the silicone circuit that constitute the solution in an exponential manner [10, 27, 29]. The number of the silicone circuit elements that increase as a result of the practical application of SOC Architecture cannot provide adequate development in power consumption and working conditions [18, 23, 24, 29–31]. The basic reason of this problem is the lack of an efficient ADA Approach in order to tolerate the DE acceptances [28] by AE [9, 32]. The Black Zone and Hook Up Delays are some of these basic problems in this topic [10, 11, 24, 33]. The problem stems from the increase of the number of the elements of the analogue circuits in an exponential manner due to the increasing DE door/element number without ensuring a development in AE Library Approach on which DE is working [11, 24, 33]. The numbers of the elements that increase in an exponential manner cause that the points that are not considered as errors in theoretical design process. But this expectation return in the form of physical quantity and energy format and as a difference in the temperature because of not being tolerated (Advantage Analog) [10, 24]. When the DE solution development is considered in basic terms, all the physical quantities that are present in the universe gain a meaning at macro level and in an analogue manner [34–36]. In the world in which we live on an analogue axis, it is observed that, firstly, analogue physical data must be converted into numerical form, and the losses and errors must be tolerated and the entry data must be transferred into digital medium in order to develop a digital solution that is proper for the targeted aim [11, 15, 33]. Then, the digital data that is transferred is processed and interpreted in a manner that is proper for the target, and the result is thus formed [11, 24, 33]. The digital result data obtained is re-produced as an analogue output by caring for the possible losses and errors [10, 24]. This defined cycle fictionalizes an extremely long, inconvenient and non-productive work intended for a simple decision-making process [34–36]. This solution approach that is defined above and that is used in today's world must be made to become more efficient, which is a necessity for the electronic science [11, 13, 33]. For the solution approach, firstly, the AE and DE are examined in terms of the design and production processes, and the results are presented in Table 1 in a comparative manner. The positive and negative sides of AE are summarized below over the data obtained [15, 37].

Advantages of Analog Circuits:

- Due to simpler structures, circuits have the lower power consumption.
- Due to simpler structures, circuits have more flexible working conditions.
- Re-programmable analog integrated circuit design tools exist.
- Lower production costs and production processes are repeatable.

Disadvantages of Analog Circuits:

- Model library construction and the mathematical approach are difficult.

(continued)

- There is a complex relationship among the circuit elements, values, and numbers.
- Some Analog Electronic components cannot be used in integrated circuits.
- There is a complex relationship between component placement and optimization, which creates a problem.
- Library approach is needed to standardize on VLSI Design.

It is suggested that the Gene Law [18], which is known by his own name, is used in the definition of the road work released by Gene Franz in 2000 with the target of reaching comparative power consumption for AE and DE by taking the developments and expectations on digital sign processing as the bases. In Gene Law, the mW/MIPS rate decreases in half in every 18 months [18]. The proposed Gene Law covers and confirms the Moore Law [34, 35]. According to Gene Law, the information is provided claiming that analogue sign processing units will have lower power consumption 20 years ahead when compared with digital signal processing units (ADC-supported) in the power consumption decrease curve for years given in Fig. 3.

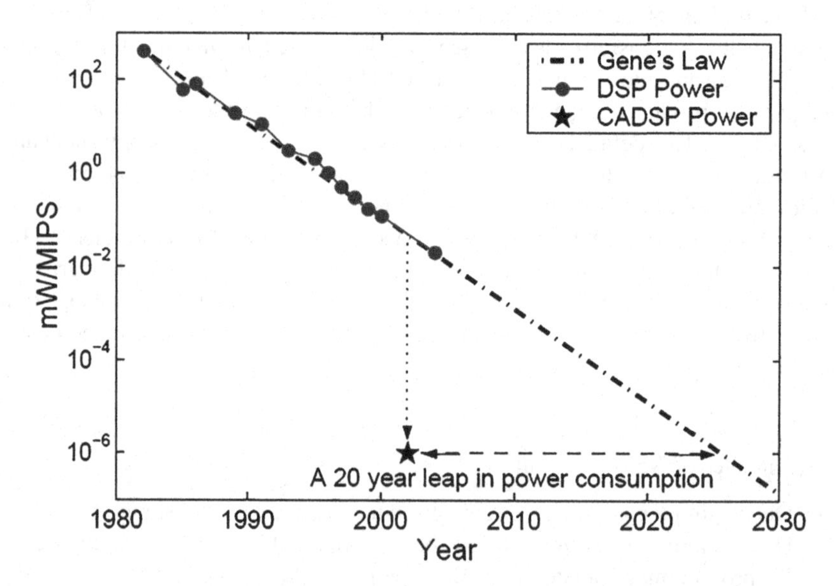

Fig. 3 Comparison of the power consumption trends in DSP [18] and analog chip developed by the CADSP team [34, 35]

The confirmation work targeted for the proposed law is confirmed practically by the CADSP team [23, 30, 31, 34, 35]. The target of less power consumption being achieved in analogue systems 20 years before the achievement of the same target in digital systems created an inclination to AE-based circuit models in many fields, mainly in the sign processing field [9, 18, 23, 30, 31].

The FPGAs were developed in order to perform modular production with fast prototyping and to allow the designers confirm practically in terms of DE development process [16, 27]. When AE is considered, it is observed that a FPGA-like approach is developed [24, 33]. The FPAA was developed for AE for the purpose of practical confirmation and modular production [20]. With the developed FPAA proposal, it is foreseen that similar automation systems may be established like in DE [33]. However, it is observed that AE includes more variables than DE, and the form of simple circuit approach is definitely not in the binary system form, and has exponentially increasing complexity [17, 22, 38, 39]. When the design process is considered in terms of calculation-based approaches without practical confirmations, it is observed that the DE Theoretical Calculation Approaches allow that real-time research may be conducted on a separate and continuous time axis [27]. It is also observed that the same situation is not valid for AE [33]. For subordinated circuit models and circuit solutions in AE, many mathematical approaches that are still used today were developed [11, 24, 28]. In the developed models, when the circuit size increases, the number of the variable-elements in the calculations and the relations between each other on the time axis also develop [13, 33]. The requirement for calculation, which shows increase, also increases the processing time in an exponential manner [10, 25, 28, 33]. For the solution of this problem, there is various software that were developed over the computer units with high processing and data capacities, which ensure that the results of analog circuits are observed with simulation calculation [40]. This software ensures that the result is calculated over the mathematical equations that are proposed and modeled for the AE that is desired to be simulated [33]. The resolution used during the calculations and the error tolerance values of the calculation models influence the result in a direct manner [13, 33]. It is observed that the simulation data and the results that are obtained by the practical testing of analogue circuits are not the same, and as the circuit size and sensitivity develop, the error rate reaches incredible levels [41]. Although the aforementioned analogue circuit design and application problems are reported, design and production engineers still continue to develop analogue circuits in an intense manner and design more productive subordinated circuit models [15, 24]. On the basis of this labor-intensive effort, there is the fact that analogue circuits provide less productive, lower power consuming, wide spectrum and high performance abilities when compared with numerical solutions [9, 15, 18, 23, 24, 30, 31, 34, 35].

4 Missing Details in the Local Visual Meaning Analogue Works

When the studies in the literature are examined, more than 20 solution models, which were developed since 1980 for the same targeted solution, are observed. These related studies presented as an overview in Table 3. When the relevant models are examined, it is observed that they brought proposals for the solution of AE circuit design problem, and for the solution of ADA design automation solutions. When the approaches developed by the solution models are examined, the following results are concluded;

- Practically, the application and confirmation of the result of the study within the automation system (AE Offset and Validation problem),
- Making the system become parallel for a higher system success (High Calculation Effort),
- Using simulation software in the design process (Pre-emulation),
- Using an artificial intelligence algorithm that is adapted for the purpose (Adaptive Artificial Intelligence), and
- Online assessment of the design result (Online Simulation for total span).

The requirements of ADA, which were determined as five items, are given in a comparative manner with solution approaches in Table 2. A new innovative solution model is proposed in Fig. 4 in the light of the problems observed in literature scan by considering the missing points determined and the innovative approaches. No other studiers were observed in the literature that proposed the comprehensive and application-focused AE fast prototyping system model solution model.

It is observed that some parts of the proposed system model are included in some studies, and results are produced and released by using the relevant parts that were examined. This information provides confirmation on the requirement of the sub-units placed in the proposed system model, and is selected by considering the system target and ADA automation system requirements. It is targeted that the system is placed in the literature as an authentic model with the joining of the model parts presented in Table 2 and by making them work together to receive results. It is considered that the process parts and steps for the proposed ADA system model being present in similar studies in the literature will not prevent the system model; on the contrary, it will confirm the success of the system. The solution approaches in the proposed system model for the analogue circuit problems are given in Table 1 (Table 3).

Table 2 Analog circuit design big data problems and local visual meaning system model approach [20]

Detected future problems	Suggestion for solution in the system model
Difficult in terms of basic library and mathematical approach to circuit design	Using the analog circuits simulation software HSPICE for containing the most recent and innovative solutions with the primary outcome analysis
A combination of numbers and values of the possible elements of the circuit element is an infinite number of elements	A growing set of research space, does higher number research with parallel computing and narrowing the focus of research space on the basis of the practical limitations of the hardware verification
Integrated in-house design and manufacturing process must be supported with the SOC approach. But some analog circuit elements cannot be taken into an integrated	Instead of analog circuit elements, which are not included in the practical verification unit, use corresponding solutions developed in FPAA
Placement and optimization approach for circuit, complex relationship between the circuit elements is a problem	With the working on practical application unit the connection matrix, more complex relationships keep out of the study
VLSI Design for the standardization, the library approach is needed	With storage obtained solutions to database, make VLSI model library

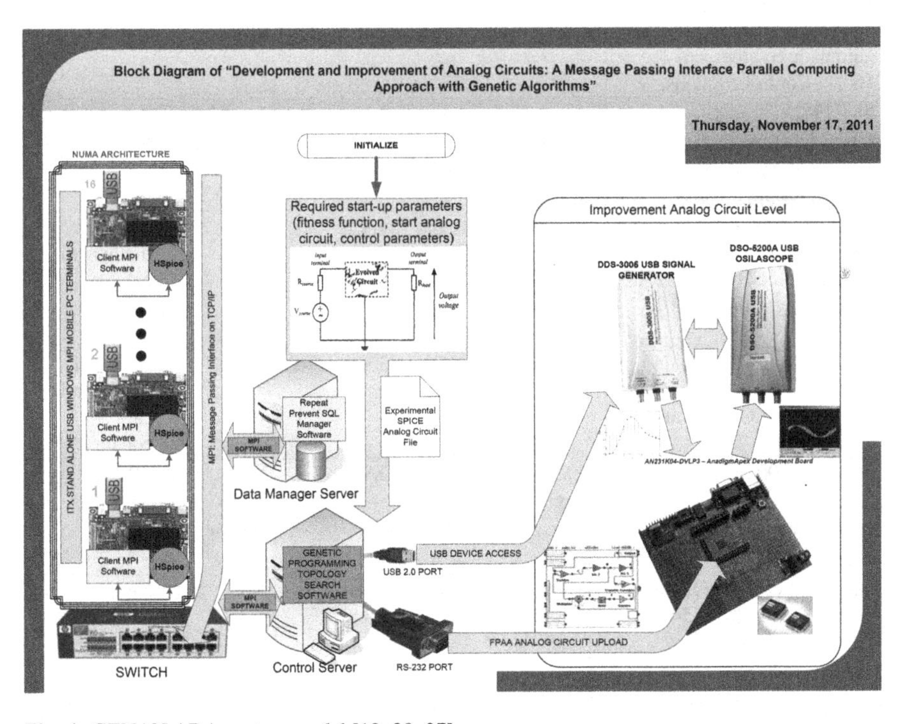

Fig. 4 GENAN ADA system model [12, 20, 37]

Table 3 Comparison table of similar studies including advanced computing

Name and year of study	The goal of the study	Artificial intelligence algorithm	Parallel processing	Circuit simulation software usage
System For Optimization of Electronic Circuits Using Genetic Algorithm, 1996 [42]	Analog operational transconductor amplifier (OTA) circuits	GA	Yes (Slave Terminal, Novell)	PSpice
Dynamically Reconfigurable Analog/Digital Hardware—Implementation Using FPGA and FPAA Technologies, 1998 [43]	Analog mixed signal circuits	MATLAB	No	No
VASE, 1999 [44]	Analog RF circuits	GA	No	SPICE
PAMA, 2000 [45]	Analog circuit design	GA	No	
EORA SOC, 2000 [46]	Analog circuits temperature responds	GA	No	SPICE
Automatic Analogue Circuit Synthesis using Genetic Algorithms, 2000 [47]	Analog filter circuits	GP and GA	No	Spice
Evolvable Hardware: On the Automatic Synthesis of Analog Control Systems, 2000 [48]	Analog MOS	GA	No	PSPICE
A Parallel Genetic Algorithm for Automated Electronic Circuit Design, 2000 [49]	Analog filter and OPAMP circuits	GA	Yes (master-slave)	SPICE
Design Methodology For Optimization of Analog Building Blocks Using Genetic Algorithms, 2001 [50]	Analog filter circuits	GA	No	No (with mathematical equation)
GA Automated Design and Synthesis of Analog Circuits with Practical Constraints, 2001 [51]	Analog filter circuits	GA	No	No (with mathematical equation)
SABLES, 2002 [52]	Analog circuit design	GA	No	SPICE
BIST, 2003 [53]	Analog circuit design	N/A	No	No
Adaptive Real-Time Systems and the FPAA, 2003 [54]	Analog filter circuits	Active gain control algorithm	No	No
PAMA, 2004 [55]	Analog circuit design	GA	No	No

(continued)

Table 3 (continued)

Name and year of study	The goal of the study	Artificial intelligence algorithm	Parallel processing	Circuit simulation software usage
Application Performance Of Elements in a Floating–Gate FPAA, 2004 [34]	Analog OTA filter circuits	Simulink (MAT-LAB)	No	RASP 2.8
CAFFELINE, 2005 [56]	Nonlinear analog circuit models	GP, NSGA-II	No	SPICE
Reconfigurable RF Circuit Design, 2005 [57]	Analog RF circuits	RF MEMS	No	No
Generic Mixed-signal Rapid Prototyping Platform, 2005 [58]	Analog B and stop filter circuits	NSGA-II	No	No
Mixed–Signal Prototyping of Embedded Systems, 2005 [59]	Analog signal process circuits	N/A	No	No
RASP, 2005 [35]	Analog OTA filter circuits	Simulink (MAT-LAB)	No	RASP 1.5
Evolutionary Design of Analog Circuits with A Uniform Design Based Multi Objective Adaptive Genetic Algorithm, 2005 [60]	Analog high pass filter	Multi-objective adaptive genetic algorithm (UMOAGA)	No	No (with mathematical equation)
An Automated Circuit Design Procedure by Means of Genetic Programming, 2005 [61]	Analog filter circuits	GP	No	SPICE
GRACE, 2006 [62]	Analog circuit design	CGP, PSO	No	No
Parallel Genetic Algorithm for SPICE Model Parameter Extraction, 2006 [63]	Analog circuit component value definitions	GA	Yes (MPI, Linux Cluster)	BSIM4 SPICE
Reconfigure Parallel Architecture for Genetic Algorithms, 2007 [64]	Digital circuit design	GA, CGP	No	No
Self-Organizing Analogue Circuit by Monte Carlo Method, 2007 [65]	Analog low pass filter	Monte Carlo method	No	Spice
On the "Evolvable Hardware" Approach to Electronic Design Invention, 2007 [66]	Quantum computer	N/A	No	N/A
RASPER, 2008 [67]	Analog filter circuits	Simulink (MAT-LAB)	No	RASP 2.7, SPICE
Steady-State and Transient Evaluation of FPAA, 2009 [68]	Analog circuit design	MLS	No	No

(continued)

Table 3 (continued)

Name and year of study	The goal of the study	Artificial intelligence algorithm	Parallel processing	Circuit simulation software usage
Development of Analog Circuit Design Automation Tool, 2009 [69]	Analog CMOS OPAMP	GA	No	Berkeley SPICE
GRASPER, 2010 [70]	Analog circuit design	Simulink (MATLAB)	No	SPICE (Sim2Spice)
Simulation and Implementation of Adaptive and Matched Filters Using FPAA Technology, 2010 [71]	Analog adaptive filter	N/A	No	Anadigm designer 2 EDA
MFPAA, 2012 [19]	Analog filter circuits design	Simulink (MATLAB)	No	Reconfigurable analog signal processor (RASP) 2.8
ANEHP-Alpha, 2008 [32]	Analog filter circuits design	Immune GA	No	N/A
LAYGEN II, 2013 [72]	Analog VLSI circuit design	NSGA-II	No	Mentor graphics' calibre

5 Pilot Application: Adaptive BPF Analog Circuit

Filters are placed in many electronic systems as basic components [73]. Traditional and fast resolutions are fictionalized in the field of analogue structures. However, adaptation of the analogue filter solutions developed for the system and solution-based requirements is an extremely difficult engineering problem. FPAA provides adaptive solutions in this topic with its dynamic concept. The basic difference between adaptive filter and classic filter circuit is the fact that the filter circuit keeps the error coefficient at minimum level by adapting itself on the bases of the requirements that will occur in necessary conditions [71]. The microprocessor in the FPAA supports the adaptation of the filter circuit developed with the controlled signal producer. The analogue circuit model used for the targeted aim is given in Fig. 5. The comparison of result-quality values are given in Fig. 6. S1-S8 Switching Elements change the cut-off frequency, and thus, the adaptive quality factor of the filter may be changed. The equations given by the FPAA producer company Anadigm Corporation are given in Eqs. (1) and (2) [74]. The switching values in the scope of fostart=10 Khz/fostop = 11 Khz value selected for targeted circuit aim are given in Table 4 [75].

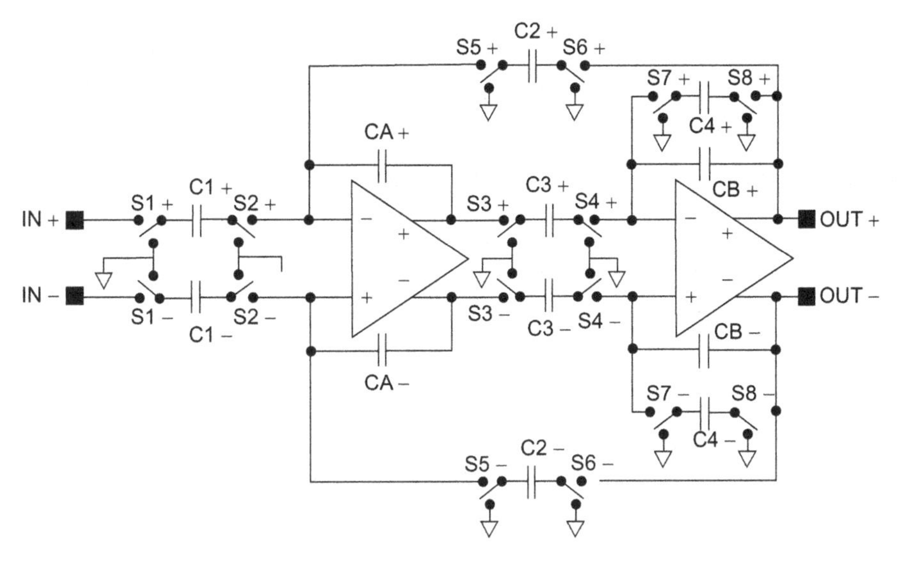

Fig. 5 General purpose adaptive filter circuit [74]

$$f_o \cong \frac{f_c}{2\pi}\sqrt{\frac{C_2\,C_3}{C_A\,C_B}} \tag{1}$$

$$Q \cong \frac{C_B}{C_4}\sqrt{\frac{C_2 C_3}{C_A C_B}} \tag{2}$$

6 Conclusions and Future Directions

There is a need for high calculation effort in the "analog circuit big data local visual meaning", like an adaptive filter design and production with lower power consumption. In addition to this, the circuit model that will be produced must be enabled to work under various conditions. The basic aim and scope of this study is to propose a solution model meaning analytics that makes it possible to elimination of latency, develop a circuit rapid model in the required qualities. Three basic targets are supported with the proposed solution model; adaptiveness, low cost, and wide usage area. As a conclusion, the developed system model fictionalizes a pilot solution model for high-capacity production models for similar aims in terms of high-capacity and critical resolution.

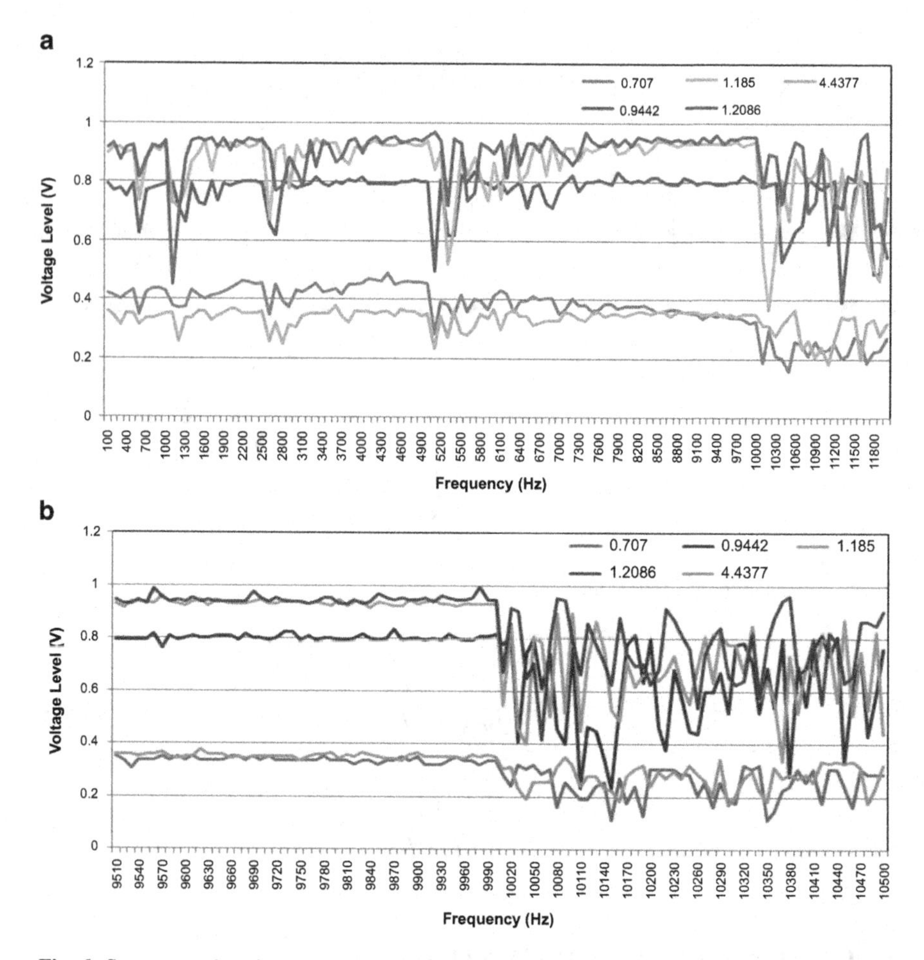

Fig. 6 Spectrum values for best quality values; (**a**) 0.1–12 KHz, (**b**) 9.51–10.5 KHz [75]

Table 4 Objective purpose switching value and quality factor

The numerical value of the S keys (ACLK = 4 Mhz)								Quality factor (Q)
C1		C3		C2		C4		
S1	S2	S3	S4	S5	S6	S7	S8	
222	221	10	10	225	225	5	5	0.707[a]
237	183	55	86	136	130	74	187	0.9442
89	59	39	224	217	136	93	206	1.1850
78	103	8	224	191	189	93	211	1.2086
30	103	8	234	25	189	232	211	4.4377

[a]FPAA fo = 10 Khz library value [74]

References

1. What was the largest dataset you analyzed/data mined? poll, KD nuggets. www.kdnuggets.com/polls/2015/largest-dataset-analyzed-data-mined.html (2015)
2. Fisher, D.: Big data exploration requires collaboration between visualization and data infrastructures. In: Proc. Workshop on Human-in-the Loop Data Analytics (HILDA), article no. 16 (2016)
3. Hellerstein, J.M., et al.: Interactive data analysis: the control project. Computer. **32**(8), 51–59 (1999)
4. Kwon, B.C., Verma, J., Haas, P.J., Demiralp, Ç.: Sampling for scalable visual analytics, visualization viewpoints. IEEE Computer Society 0272–1716/17, January/February, pp. 100–108 (2017)
5. Turner, V., Gantz, J., Reinsel, D., Minton, S.: The digital universe of opportunities: rich data and the increase value of the internet of things. IDC Anal. Future (2014)
6. Hu, H., Wen, Y., Chua, T.S., Li, X.: Toward scalable systems for big data analytics: a technology tutorial. IEEE Access. **2**, 652–687 (2014)
7. Wu, X., Zhu, X., Wu, G.Q., Ding, W.: Data mining with big data. IEEE Trans. Knowl. Data Eng. **26**(1), 97–107 (2014)
8. Liu, Z., Heer, J.: The effects of interactive latency on exploratory visual analysis. IEEE Trans. Visual Comput. Graphics. **20**, 2122–2131 (2014)
9. Middlebrook, R.D.: Analog design: the academic view, technology trends, electronic, engineering times, pp. 87–88, December 17 (1990)
10. Toumazou, C., Moschytz, G., Gilbert, B.: Trade-Offs in Analog Circuit Design: The Designer's Companion. Kluwer Academic, Dordrecht (2002.), ISBN 1-4020-7037-3
11. Zebulum, R.S., Pacheco, M.A.C., Vellasco, M.M.: Evolutionary Electronics Automatic Design of Electronic Circuits and Systems by Genetic Algorithms, pp. 3–6. CRC, Boca Raton (2001), ISBN 0-8493-0865-8
12. Aksu, O.: Investigation of evolution hardware studies on analog circuits for designing healthcare systems. In: The Society for Design and Process Science (SDPS), Conference, Orlando, 4–6 December (2016)
13. Quentin, K.G., et al.: Tools for computer-aided design of multigigahertz supeconducting digital circuits. IEEE Trans. Appl. Supercond. **9**(1), 18–38 (1999)
14. Williams, J.: Analog circuit design: art, science and personalities, ISBN 978–0750696401, pp. 15–16 (1991)
15. Sarpeshkar, R.: Analog versus digital: extrapolating from electronics to neurobiology. Neural Comput. **10**, 1601–1638 (1998)
16. Aggarwal, V., Berggren, K., O'Reilly, U.: On the "evolvable hardware" approach to electronic design invention. In: Proceedings of the 2007 WEAH, IEEE Workshop on Evolvable and Adaptive Hardware (2007)
17. Koza, R.J., et al.: Evolutionary design of analog electrical circuits using genetic programming. In: Adaptive Computing in Design and Manufacture Conference (ACDM-98) (1998)
18. Franz, G.: Digital signal processor trends. IEEE Micro. **20**(6), 52–59 (2000)
19. Schlottmann, C.R., Abramson, D., Hasler, P.E.: A MITE-based translinear FPAA. IEEE Trans. Very Large Scale Integr. (VLSI) Syst. **20**(1), 1–9 (2012)
20. Aksu, O., Kalinli, A., Tanik, M.: Development and improvement of analog circuit design: an adjustable analog signal generation circuit approach. In: SDPS 2012 Conference, Berlin, 10–15 June (2012)
21. Platonov, A.: Analog transmission more efficient than digital: can it be and when? In: 2014 International Conference on IEEE Signals and Electronic Systems (ICSES), pp. 1–4, 11–13 September (2014)
22. Koza, R.J., et al.: Automatic creation of computer programs for designing electrical circuits using genetic programming. In: Computational Intelligence and Software Engineering. World Scientific Publishing, Singapore (1997)

23. Ellis, R., Yoo, H., Graham, D., Hasler, P., Anderson, D.: A continuous-time speech enhancement front-end for microphone inputs. In: Proceedings of the IEEE International Symposium on Circuits and Systems, vol. 2, pp. II–728–II–731, Phoenix, AZ (2002)
24. Balkir, S., Dundar, G., Ogrenci, S.: Analog VLSI Design Automation. CRC, Boca Raton (2003.), ISBN:978-0-8493-1090-4
25. Glelen, G., Sansen, W.: Symbolic analysis for automated design of analog integrated circuits. In: The Springer International Series in Engineering and Computer Science, ISBN 978-0792391616 (1991)
26. Nair, B.S.: Digital electronics and logic design. PHI Learning, Delhi. part 4-2,4-3, ISBN 978-8120319561 (2002)
27. Hornby, G.S., Sekanina, L., Haddow, P.C.: Evolvable systems: from biology to hardware: In: 8th International Conference, ICES 2008, Prague, Czech Republic, September 21–24 (2008)
28. Dobkin, B., Hamburger, J.: Analog circuit design, Newnes, vol. 3, pp. 38, ISBN 978-0128000014 (2014)
29. Eick, M., Graeb, H.E.: MARS: matching-driven analog sizing. IEEE Trans. Comput. Aided Des. Integr. Circuits Syst. **31**(8), 1145–1158 (2012)
30. Hasler, P., Smith, P., Ellis, R., Graham, D., Anderson, D.V.: Biologically inspired auditory sensing system interfaces on a chip. In: 2002 IEEE Sensors Conference, Orlando, FL (2002)
31. Smith, P.D., Kucic, M., Ellis, R., Hasler, P., Anderson, D.V.: Mel–frequency cepstrum encoding in analog floating-gate circuitry. In: Proceedings of the International Symposium on Circuits and Systems, pp. IV–671–IV–674, Phoenix, AZ (2002)
32. Zhang, W., Li, Y., Liu, N.: An online evolvable Chebyshev filter based on immune genetic algorithm. In: Proceedings of the International Multi Conference of Engineers and Computer Scientists IMECS, Hong Kong, (19–21) March (2008)
33. Malcher, A., Falkowski, P.: Analog reconfigurable circuits. Int. J. Electron. Telecommun. **60**, 15–26 (2014)
34. Hall, T.S., Twigg, C.M., Hasler, P., Anderson, D.V.: Application performance of elements in a floating-gate FPAA, In: Proceedings of the 2004 International Symposium on Circuits and Systems, 2004. ISCAS '04, vol. 2, pp. II-589-92 (2004)
35. Hall, T.S., Twigg, C.M., Gray, J.D., Hasler, P., Anderson, D.V.: Large-scale field-programmable analog arrays for analog signal processing. IEEE Trans. Circuits Syst. Regul. Pap. **52**(11), 2298–2307 (2005)
36. Lohn, J.D., Colombano, S.P.: A circuit representation technique for automated circuit design. IEEE Trans. Evol. Comput. **3**(3), 205–219 (1999)
37. Aksu, O., Kalınlı, A.: Development and improvement of analog circuit design: a message passing interface parallel computing approach with genetic algorithms. Soc. Des. Process Sci. (SDPS). **14**(3), 37–52 (2010)
38. Bennett III, H.F., et al.: Automatic Synthesis, Placement and Routing of an Amplifier Circuit by Means of Genetic Programming. FX Palo Alto Laboratory, Palo Alto, CA (2000)
39. Koza, J.R., et al.: Automated design of both the topology and sizing of analog electrical circuits using genetic programming. In: Gero, J.S., Sudweeks, F. (eds.) Artificial Intelligence in Design '96, pp. 151–170. Kluwer Academic, Dordrecht (1996)
40. Grimbleby, J.B.: Automatic analogue circuit synthesis using genetic algorithms. In: IEEE Proceedings Circuits Devices and Systems, vol. 147, no. 6 (2000)
41. Brambilla, A., D'Amore, D.: The simulation errors introduced by the SPICE transient analysis. IEEE Trans. Circuits Syst. I, Fundam. Theory Appl. **40**(1), 57–60 (1993)
42. Wojcikowski, M., Glinianowicz, J., Bialko, M.: System for optimisation of electronic circuits using genetic algorithm. In: Proceedings of the Third IEEE International Conference on Electronics, Circuits, and Systems, ICECS '96, vol. 1, pp. 247–250 (1996)
43. Reiser, C., et al.: Dynamically reconfigurable analog/digital hardware-implementation using FPGA and FPAA technologies. J. Circuits Syst. Comput. World Sci Pub (1998)
44. Ganesan, S., Vemuri, R.: A methodology for rapid prototyping of analog systems. In: 1999 (ICCD '99) International Conference on Computer Design, pp. 482–488 (1999)

45. Zebulum, R., Sinohara, H., Vellasco, M., Santini, C., Pacheco, M., Szwarcman, M.: A reconfigurable platform for the automatic synthesis of analog circuits. In: The Second NASA/DoD Workshop on Evolvable Hardware, Proceedings, pp. 91–98 (2000)
46. Stoica, A., Keymeulen, D., Zebulum, R., Thakoor, A., Daud, T., Klimeck, Y., Tawel, R., Duong, V.: Evolution of analog circuits on field programmable transistor arrays. In: The Second NASA/DoD Workshop on Evolvable Hardware, Proceedings, pp. 99–108 (2000)
47. Grimbleby, J.B.: Automatic analogue circuit synthesis using genetic algorithms. IEEE Proc. Circuits Devices Syst. **147**(6), 319–323 (2000)
48. Zebulum, R.S., Vellasco, M., Pacheco, M.A., Sinohara, H.T.: Evolvable hardware: on the automatic synthesis of analog control systems. In: Aerospace Conference Proceedings, IEEE, vol. 5, pp. 451–463 (2000)
49. Long, J.D., et al.: A parallel genetic algorithm for automated electronic circuit design (2000)
50. Paulino, N., Goes, J., Steiger-Garcao, A.: Design methodology for optimization of analog building blocks using genetic algorithms. In: The 2001 IEEE International Symposium on Circuits and Systems, ISCAS 2001, vol. 5, pp. 435–438 (2001)
51. Goh, C., Li, Y.: GA Automated design and synthesis of analog circuits with practical constraints. In: Proceedings of the 2001 Congress on Evolutionary Computation, vol. 1, pp. 170–177 (2001)
52. Stoica, G., Xin, R.S., Zebulum, M.I., Ferguson, D.: Evolution-based automated reconfiguration of field programmable analog devices. In: IEEE International Conference on Field-Programmable Technology (FPT), pp. 403–406 (2002)
53. Sanahuja, R., Barcons, V., Balado, L., Figueras, J.: Experimental test bench for mixed-signal circuits based on FPAA devices. In: Conference on Design of Circuits and Integrated Systems (DCIS), pp. 344–349 (2003)
54. Colsell, S., Edwards, R.: Adaptive real-time systems and the FPAA. In: Field Programmable Logic and Application, Lecture Notes in Computer Science, vol. 2778, pp. 944–947. Springer Berlin, Heidelberg (2003.), 978-3-540-40822-2
55. Santini, C.C., Amaral, J.F.M., Pacheco, M.A.C., Tanscheit, R.: Evolvability and reconfigurability. In: IEEE International Conference Field-Programmable Technology, Proceedings, pp. 105–112 (2004)
56. McConaghy, T., Eeckelaert, T., Gielen, G.: CAFFEINE: template-free symbolic model generation of analog circuits via canonical form functions and genetic programming. In: Design, Automation and Test in Europe, vol. 2, pp. 1082–1087 (2005)
57. Okada, K., Yoshihara, Y., Sugawara, H., Masu, K.: Reconfigurable RF circuit design. In: 18th Asia and South Pacific Design Automation Conference, ASP-DAC, pp. 683–686 (2005)
58. Ghali, K., Dorie, L., Hammami, O.: Dynamically reconfigurable analog circuit design automation through multiobjective optimization and direct execution. In: 12th IEEE International Conference on Electronics, Circuits and Systems, ICECS 2005, pp. 1–4 (2005)
59. Hall, T., Twigg, C.: Field-programmable analog arrays enable mixed-signal prototyping of embedded systems. In: 48th Midwest Symposium on Circuits and Systems, vol. 1, pp. 83–86 (2005)
60. Zhao, S., Jiao, L., Zhao, J., Wang, Y.: Evolutionary design of analog circuits with a uniform-design based multi-objective adaptive genetic algorithm. In: NASA/DoD Conference on Evolvable Hardware, pp. 26–29 (2005)
61. Jin'no K.: An automated circuit design procedure by means of genetic programming. In: 2005 International Symposium on Nonlinear Theory and its Applications (NOLTA2005) Bruges, Belgium, pp. 194–197, October 18–21 (2005)
62. Terry, M. A., Marcus, J., Farrell, M., Aggarwal, V., O'Reilly, U.: GRACE: generative robust analog circuit exploration. In: Applications of Evolutionary Computing EvoWorkshops2006 EvoBIO EvoCOMNET EvoHOT EvoIASP EvoInteraction EvoMUSART EvoSTOC, vol. 3907, pp. 332–343 (2006)
63. Li, Y., Cho, Y.: Parallel genetic algorithm for SPICE model parameter extraction. In: Parallel and Distributed Processing Symposium, IPDPS 20th International, pp. 8, 25–29 April (2006)

64. Ferlin, E.P., Lopes, H.S., Lima, C.R.E., Cichaczewski, E.: Reconfigure parallel architecture for genetic algorithms: application to the synthesis of digital circuits. In: Third International Workshop, ARC 2007, Brazil, pp. 326–336, March 27–29 (2007)
65. Gyorok, G.: Self organizing analogue circuit by Monte Carlo method. In: LINDI International Symposium on Logistics and Industrial Informatics, pp. 37–40, 13–15 September (2007)
66. Aggarwal, V., Berggren, K., O'Reilly, U.: On the evolvable hardware approach to electronic design invention. In: 2007 WEAH IEEE Workshop on Evolvable and Adaptive Hardware, pp. 46–54, 1–5 April (2007)
67. Petre, C., Schlottmann, C., Hasler, P.: Automated conversion of Simulink designs to analog hardware on an FPAA. In: IEEE International Symposium on Circuits and Systems, ISCAS 2008, pp. 500–503, 18–21 May (2008)
68. Potirakis, S.M., Deli, J., Rangoussi, M.: steady-state and transient evaluation of FPAA implemented analog filters using a MLS system analyzer. In: 16th International Conference on Systems, Signals and Image Processing, IWSSIP 2009, pp. 1–8, 18–20 June (2009)
69. Krishnamurthy, V., Kim, B.: Development of analog circuit design automation tool, Southeastcon, SOUTHEASTCON '09, IEEE, pp. 236–241, 5–8 March (2009)
70. Koziol, S., et al.: Hardware and software infrastructure for a family of floating-gate based FPAAs. In: Proceedings of 2010 IEEE International Symposium on Circuits and Systems (ISCAS), pp. 2794–2797, May 30–June 2 (2010)
71. Visan, D.A., Lita, I., Jurian, M., Cioc, I.B.: Simulation and implementation of adaptive and matched filters using FPAA technology. In: 2010 IEEE 16th International Symposium for Design and Technology in Electronic Packaging (SIITME), pp. 177–180, 23–26 September (2010)
72. Martins, R., Lourenco, N., Horta, N.: LAYGEN II—automatic layout generation of analog integrated circuits. IEEE Trans. Comput. Aided Des. Integr. Circuits Syst. **32**(11), 1641–1654 (2013)
73. Haji Ali, M.S., Shaker, M.M., Salih, T.A.: Design and implementation of a dynamic analog matched filter using FPAA technology. IEEE J. Solid State Circuits. **23**(6), 1298–1308 (2008)
74. Anadigm, Inc. http://www.anadigm.com
75. Aksu, O.: Investigation of evolution hardware studies on analog circuits for rapid prototyping and proposal a new model, ELECO, pp. 147–155, Bursa (2016)

Transdisciplinary Benefits of Convergence in Big Data Analytics

U. John Tanik and Darrell Fielder

Abstract Big Data applications can benefit from transdisciplinary convergence spanning multiple domains and opportunity areas. These big data applications in areas as diverse as healthcare, energy, and business must now process a tremendous volume of data and from a variety of data sources. New ways to process and store information beyond the standard three V's of information characterization defined by volume, variety, and velocity are needed, to also include data variability and veracity. As such, volume and variety of data types alone pose significant challenges to timely and accurate analysis by human as well as machine operators. Although big data applications can instantiate human logic into executable code, process the volume of data quickly, and make correlations across the variety of data, leading to better analysis and predictive capabilities, the resulting conclusions are delimited by other key factors imposed by data velocity that limits immediate inferencing, data variability that limits data consistency, and data veracity that limits data quality. As these challenges are resolved in other disciplinary domains, the spillover benefits can be profound when transdisciplinary advances in one particular domain impact other domain application areas by way of convergence.

1 Introduction

Big data applications can benefit from transdisciplinary convergence spanning multiple domains and opportunity areas. These big data applications in areas as diverse as healthcare, energy, and business must now process a tremendous volume of data (e.g. gigabytes, terabytes, and exabytes) and from a variety of data sources (e.g. customer, inventory, environmental, network, social media, monitors, and sensors). Furthermore, according to IBM (International Business Machines), at the projected growth rates, the volume of healthcare data alone will soon be zettabyte

U.J. Tanik (✉)
Department of Computer Science, Texas A&M University-Commerce, Commerce, TX, USA
e-mail: john.tanik@tamuc.edu

D. Fielder
University of Alabama at Birmingham, Birmingham, AL, USA
e-mail: dfield19@uab.edu

S.C. Suh, T. Anthony (eds.), *Big Data and Visual Analytics*,
https://doi.org/10.1007/978-3-319-63917-8_9

and yottabyte scale [1] requiring new ways to process and store information beyond the standard three V's of information characterization defined by volume, variety, and velocity, to also include variability, and veracity. As such, volume and variety of data types alone pose significant challenges to timely and accurate analysis by human as well as machine operators. Although Big Data applications can instantiate human logic into executable code, process the volume of data quickly, and make correlations across the variety of data, leading to better analysis and predictive capabilities, the resulting conclusions are delimited by other key factors imposed by data velocity that limits immediate inferencing, data variability that limits data consistency, and data veracity that limits data quality. As these challenges are resolved in other disciplinary domains, the spillover benefits can be profound when transdisciplinary advances in one particular domain impact other domain application areas by way of convergence.

SDPS (Society for Design and Process Science) was formed in 1995 as a visionary society that recognized the importance of transdisciplinary notions impacting the world in the near future [2], and later confirmed in 2014 by the National Research Council with the publication *Convergence: facilitating transdisciplinary integration of life sciences, physical sciences, engineering, and beyond* [3]. As in the paper *Transdisciplinary Healthcare Engineering: Implementing a Transdisciplinary Education and Research Program*, it is important to establish the main differences between the notions of disciplinarity, which are briefly defined in Table 1 [4]. For instance, the discipline of healthcare engineering can utilize the properties of transdisciplinarity to form an information bridge to other disciplines enabling convergent applications to be developed. Thus, transdisciplinarity "joins, integrates, and/or crosses disciplines by dissolving disciplinary boundaries" enabling advances in healthcare engineering to cascade to other disciplines without significant barriers.

This chapter will also introduce the value of convergence as a key facilitator of the transdisciplinary pollination of big data techniques, technologies, platforms, and knowledge. As defined by Fielder et al., convergence allows a multi-disciplinary approach to problem solving that requires the fusion of technology, devices, knowledge, process, and people from multiple engineering disciplines [5] and across multiple data boundaries. In fact, industry analysts at MIT (Massachusetts Institute of Technology) already recognized that convergence will be the emerging paradigm to spur medical research in the future [6]. This trend was later confirmed by our broader literature survey and analysis presented at recent SDPS/IEEE conferences [5, 7–9].

Table 1 Trandisciplinarity bridging disciplinary barriers [4]

Approach	Short description
Multidisciplinarity	Joins together disciplines without integration
Interdisciplinarity	Integrates disciplines without dissolving disciplinary boundaries
Crossdisciplinarity	Crosses disciplinary boundaries to explain one subject in terms of another
Trandisciplinarity	Joins, integrates, and/or crosses disciplines by dissolving disciplinary boundaries

As previously noted, the volume and variety of data continues to grow, and other key characteristic indicators are being added to describe the explosion of unstructured data in terms of the five V's. This expansion portends that advancements in other application domains will also benefit from the transdisciplinary information bridge. The introduction and discussion of relevant big data applications in health-care, energy and business disciplines will describe the convergence implications and transdisciplinary insight gained.

2 Background and Definition of Big Data

2.1 Big Data Origin

There are many definitions of what constitutes "big data" and differing opinions about the origin of the term. In the paper *A Personal Perspective on the Origin(s) and Development of Big Data: The Phenomenon, the Term, and the Discipline*, Frances Diebold investigated the origin of the term "big data" and determined that some of the first references to big data were made by Charles Tilly as early as 1984, Eric Larson in a 1989 Washington Post article, and also in a PR Newswire, Inc. article in 1996 [10]. However, Diebold indicated that interpretation of the term "big data" appear inconsistent with the current usage and understanding of the term. Diebold concludes his research by indicating the term probably originated in lunch-table conversations at Silicon Graphics, Inc. (SGI) in the mid-1990s with John Mashey figuring prominently [37]. Other sources mention a publication by Weiss in 1998 as one of the first documents to mention big data. Lastly, according to SAS, the concept of big data gained momentum in the early 2000s when industry analyst Doug Laney articulated the now-mainstream definition of big data [11].

2.2 Big Data Definition

In addition to our understanding of the origin of the term big data, the definition of big data continues to evolve. For instance, the Merriam-Webster dictionary defines big data as an accumulation of data that is too large and complex for processing by traditional database management tools [12]. Industry analyst Doug Laney articulated the now mainstream definition of big data as the three V's—volume, velocity, and variety [11]. Big data is also defined as a holistic information management strategy that includes and integrates many new types of data and data management alongside traditional data [13]. Additionally, big data contains four V's—volume, velocity, variety, and value [13]. In their depiction of big data, IBM included veracity or uncertainty as a key element of big data [1]. Furthermore, big data is defined as high-volume, high-velocity, and/or high-variety information assets that demand cost-effective, innovative forms of information processing that enable enhanced insight, decision making, and process automation [14]. Big data is also described as data that is too big to be processed on one server, too fast moving to be sequestered in a data warehouse, or too unstructured to fit into a conventional database. Likewise, big data can be defined as a collection of very huge datasets with great diversity of types so that it becomes difficult to process by using state-of-the-art data processing approaches or traditional data processing platforms [15].

As the origin and definitions indicate, the influx of global data has been overwhelming, complex, and exponentially increasing for at least a couple of decades. However, regardless of the origin and evolution of the definition, the variety of data (image data, text data, voice data, location data, temperature data, body biometric data), volume of data (gigabytes, terabytes, and exabytes), and velocity of data (speed requirements) now pose serious challenges for institutions and companies while providing many valuable opportunities to gain unparalleled insights.

2.3 An Early Application of Big Data Approach

Data warehouses represent one example of early attempts in the development of big data approaches to collect, analyze, and gain insight from vast quantities of data. Considered by many to be the father of data warehousing, Bill Inmon first began to discuss the principles around the data warehouse and even coined the term in the 1970s [16]. In 1992, Inmon published *Building the Data Warehouse*, one of the seminal volumes of the industry. Currently in its fourth edition, the book continues to be an important part of any data professional's library with a fine-tuned mix of theoretical background and real-world examples [16].

In the 1988 article titled *An Architecture for a Business Information System*, IBM indicated the transaction-processing environment in which companies maintain their operational databases was the original target for computerization and is now well understood [17]. On the other hand, access to company information on a large scale by an end user for reporting and data analysis is relatively new. Within IBM, the computerization of informational systems is progressing, driven by business needs and by the availability of improved tools for accessing the company data.

Traditionally, decision support systems are used to obtain information from a limited amount of data to support the decision-making process [18]. However, such decision support systems have difficulty dealing with complex, multiple data sources that are typically found in large organizations. Although a data warehouse is a technology that structures data to simplify the phases of the decision support process, challenges occur when dealing with unstructured data.

Prior to data warehousing, data was often organized and stored for utilization by a single application. For example, traditional business applications such as billing and inventory were developed to perform a single function using their individual data stores. In the traditional approach, connections between the different applications and data stores were minimal though the applications shared many common features. This lack of integration made it difficult to generate cross-discipline or cross-platform insight from the comprehensive data suite available in order to use that insight to predict enterprise-wide future behavior or outcomes. Similar to pure mathematics, data analytics is a universal language that can bridge the gap between different disciplines and different platforms by interpreting the real world by measurement and analysis. Likewise, big data applications allow the expeditious processing of both structured and unstructured data for real-time decision-making, trend analysis, and future projections.

3 Disciplinary Examples of Big Data Applications

3.1 Big Data in the Healthcare Discipline

Healthcare costs are continuing to increase, outpacing inflation and other financial indicators. Healthcare now consumes 17.8% of the GDP (gross domestic product) of the Unites States [19]. The cost of treating chronic illnesses such as cancer, diabetes, heart disease, and high blood pressure is also rising. These chronic illnesses comprise a major portion of healthcare expenditures. For instance, the cost of treating cancer was estimated at $157 Billion in 2010 dollars [20], the cost of treating diabetes was $245 Billion in 2012 [21], and the cost of heart disease and stroke was $315.4 Billion in 2010 [22]. However, chronic diseases and conditions—such as heart disease, stroke, cancer, type 2 diabetes, obesity, and arthritis—are also among the most common, costly, and preventable of all health problems [23]. These chronic illnesses are affected by a convergence of complex factors including environment and lifestyle. An additional factor affecting the cost of healthcare is the increase in the elderly population defined as those aged 65 years and older. It is estimated the elderly population of the U.S will be 21.7% by Year 2040 [24]. The elderly population will require the management and treatment of many age-related chronic conditions such as heart disease, high-blood pressure, diabetes, and cancer [8]. Big data will be a key element in determining the origin of the chronic illness, devising the most effective treatment option, and positively influencing future healthcare expenditures.

With big data applications, lifestyle metrics such as food and fluid intake, blood pressure measurements, pulse rate, body temperature, miles walked, humidity, indoor and outdoor temperature, and other values can be collected, correlated, and subsequently analyzed. This cross discipline and cross platform analysis provides valuable insight into the complete healthcare environment and can be utilized to more effectively determine the root cause of illness. Big data applications in the new convergent healthcare paradigm allow delivery of healthcare services to be aligned to, and with, the participant's lifestyle [8]. This alignment leads to improved medication adherence and improved communication between the doctor and patient, both of which lead to reduced costs [8].

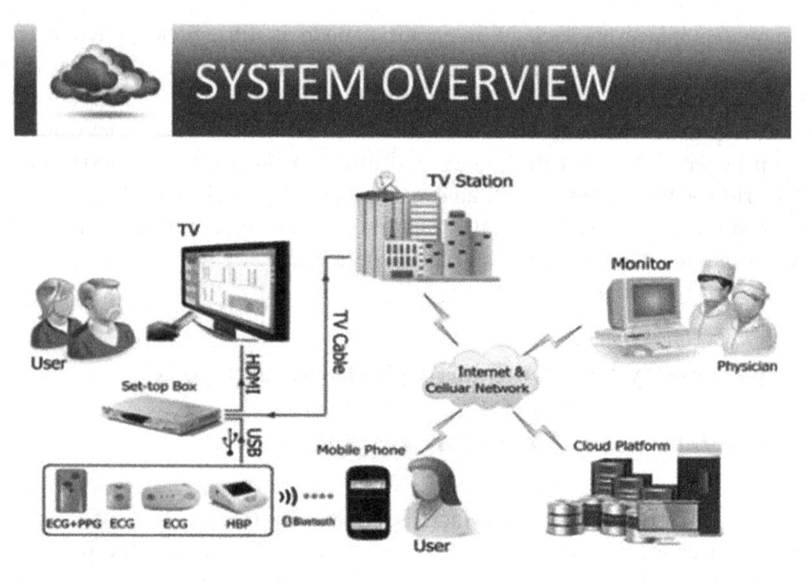

Fig. 1 Ubiquitous mobile healthcare delivery process [25]

Figure 1 provides a depiction of some key elements of a typical "ubiquitous" healthcare system. Sensing/monitoring technology and big data are vital to the delivery of healthcare in this ubiquitous, anytime, and anywhere environment. Given the different platforms, networks, and devices, big data will be an important element of the new healthcare delivery paradigm. As depicted in Fig. 1, a multitude of data values such as blood pressure and heart rate can be captured from different monitoring devices. Although not depicted, other data values such as food and fluid intake along with activity metrics (miles walked, miles run, and stairs climbed) can also be part of the solution. The captured data can be transmitted to a cloud-based application for organization and storage. Physicians can utilize the cloud-based healthcare application to access and analyze the comprehensive set of the patient's data. This enables the physician to make the best diagnosis, determine an appropriate treatment regimen, and also identify any required lifestyle adjustments. Researchers will have access to a large amount of data (individual, summary, environmental, current, and historical) from many sources. The *convergence* of data along with the ability to correlate from multiple sources will allow researchers to more quickly determine the root cause of disease and implement processes to cure, treat, and prevent future recurrence of illness.

As documented in other research articles, the demand for an extensive and mobile healthcare delivery system that is accessible at any time or from any location is accelerating [8]. This heightened interest is due to a number of factors including the rising cost of healthcare, the cost of chronic illness, growth in the elderly population, desire for more home-based monitoring, and the availability of mobile technology and the associated advanced communication network infrastructure.

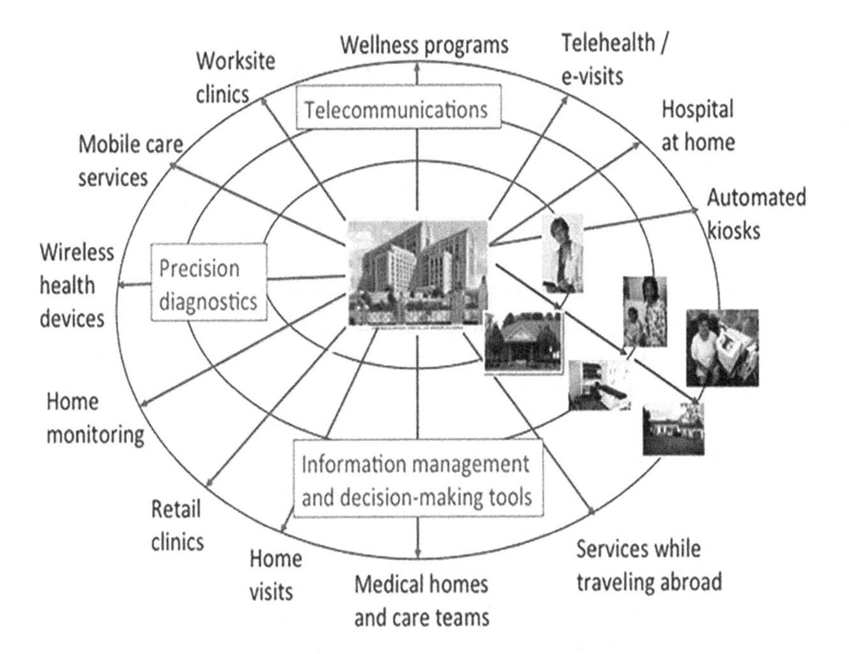

Fig. 2 Healthcare delivery process—disruptive business model [26]

Mobile healthcare delivery is both an enabler of more efficient and timely delivery of services and also a disruptor to the existing healthcare business model. As depicted in Fig. 2, the traditional healthcare delivery process must be restructured to interact with the participant across a multitude of settings and platforms. Valuable data will be available from all of these settings and platforms such as wireless health devices (fitness activity trackers), home monitoring devices (blood pressure monitors, blood glucose monitors, pulse rate monitors), workplace sites, and kiosks. This healthcare data changes quickly (velocity), differs by device (variability), and creates a huge amount of information through minute, hourly, daily, and monthly readings (volume). Big data applications will allow the data to be analyzed and correlated across the multitude of mobile devices and environments. These applications will provide a holistic view of the participant environment, enable more insight into the root cause of the illness, and devise an effective treatment regimen. The mobile delivery process also enables real-time medical decisions and efficient dispatch of valuable medical services to participants with the greatest need. Finally, mobile healthcare delivery and big data technologies allow the provision of healthcare services to be aligned to, and with, the participant's lifestyle which may lead to improved medication adherence, improved communication between doctor and patient, in addition to reduced healthcare costs.

IBM's Watson Health is an example of a specific big data application in healthcare. IBM estimates that 80% of health data is invisible to current systems because it is unstructured [27]. IBM also indicates the volume of healthcare data has recently reached 150 Exabytes. At the projected growth rates, the volume of healthcare data will soon be zettabyte and yottabyte scale. Consequently, that's enough data to fill a stack of DVDs that would stretch from Earth to Mars [27]. By integrating, collecting, and analyzing all of this big healthcare data from across various platforms and devices, Watson Health may help clinicians improve care along the healthcare spectrum—from *proactive and preventive interventions* to the *ongoing treatment* of chronic conditions, to high-touch and *multidisciplinary team care*.

However, even with the many possibilities afforded by big data in healthcare, some challenges persist. In the study "Building a Better Delivery System: A New Engineering/Health Care Partnership," the Committee on Engineering and the Health Care System from the Institute of Medicine (IOM) and National Academy of Engineering (NAE) states that a divide exists between disciplines that must be bridged [4]. We believe this chasm can be bridged by implementing a transdisciplinary approach to healthcare engineering education. Table 2 contains information from the report.

Finally, security of health data in cloud-based and other big data applications is paramount. Health information is utilized for important decisions including insurability, employment eligibility, and other key business decisions. This data must be protected and only accessed by those approved by the patient. Summary data that is not specific to a particular patient can be made available to researchers. Big data applications, combined with appropriate security, will allow the many benefits available from mobile and convergent processes to be realized and help reduce the cost of healthcare delivery.

Table 2 Quote illustrating transdisciplinary paradigm [4]

Study: "Building a better delivery system: A New engineering/health care partnership"
Source: Based on committee on engineering and the health care system from the Institute of Medicine (IOM) and National Academy of Engineering (NAE)
Quote: "...In a joint effort between the National Academy of Engineering and the Institute of Medicine, this book attempts to bridge the knowledge/awareness divide separating health care professionals from their potential partners in systems engineering and related disciplines. The goal of this partnership is to transform the U.S. health care sector from an underperforming conglomerate of independent entities (individual practitioners, small group practices, clinics, hospitals, pharmacies, community health centers et al.) into a high performance "system" in which every participating unit recognizes its dependence and influence on every other unit. By providing both a framework and action plan for a systems approach to health care delivery based on a partnership between engineers and health care professionals, Building a Better Delivery System describes opportunities and challenges to harness the power of systems-engineering tools, information technologies and complementary knowledge in social sciences, cognitive sciences and business/management to advance the U.S. health care system..."

3.2 *Big Data in the Energy Discipline*

Energy expenditures represent a significant portion of the GDP of the United States. According to the U.S. Energy Information Administration, energy expenditures as a percentage of GDP comprised 8.7% in 2006, 8.8% in 2007, 9.9% in 2008, 7.6% in 2009, and 8.3% in 2010 [28]. Additionally, whether the energy source is fossil fuel-based, nuclear or renewable, the cost of operation and maintenance (O&M) forms an important part of a power plant's business case [29]. Operations and maintenance represents those activities and functions required to keep the thousands of pieces of equipment and devices in proper functioning condition. They include day-to-day preventative and corrective maintenance, labor costs, asset and site management, maintaining health and safety, and a host of other important tasks. O&M costs can range from $20/kW produced using a Gas turbine up to $198/kW produced using Nuclear, with other forms available such as Coal, Solar, and Wind.. The scale of the GDP number and the O&M costs provide valuable information regarding the potential savings that can be derived from improvements driven by the utilization of big data.

Electric power stations consist of thousands of pieces of equipment (boilers, scrubbers, generators, turbines, solar panels, piping, etc.) that process fuel, capture emissions and other by-products, and generate electrical energy for distribution to millions of industrial, commercial, and residential customers. Each of these pieces of equipment has lifecycle requirements for periodic maintenance, repair, or replacement. Lack of maintenance, repair, or replacement may compromise the equipment, lead to failure of the equipment, and create unexpected offline periods for the power plant. In turn, this plant downtime affects the power company's ability to supply electrical energy to its customers and stability of the overall grid. The electrical grid itself also consists of thousands of pieces of equipment and devices, and thousands of miles of conductor lines. Likewise, these critical components also have lifecycle requirements for inspection, maintenance, and replacement. Figure 3 provides a depiction of an electricity delivery system. As depicted, electricity delivery to customers is a complex process beginning at the power station, crossing hundreds of miles of conductor lines, and is dependent on thousands of pieces of equipment and devices at both the power station and many points along the grid. Energy sources utilized to generate electricity at the power station are acquired from many dispersed locations, dispatched using centralized control systems, and delivered to end users residing in many disparate locations [5]. The process is complex and can benefit greatly from advancements in big data solutions.

Predictive analytic applications utilize big data and have the capability to collect data from the thousands of devices and equipment involved in the electricity generation and delivery process. Additionally, they can also collect data from the millions of devices located in commercial, residential, and industrial installations. The data utilized in predictive analytics applications will allow the power companies to more effectively determine the likelihood of failure of a piece of equipment and take *proactive* steps to prevent that failure. This big data analysis also allows

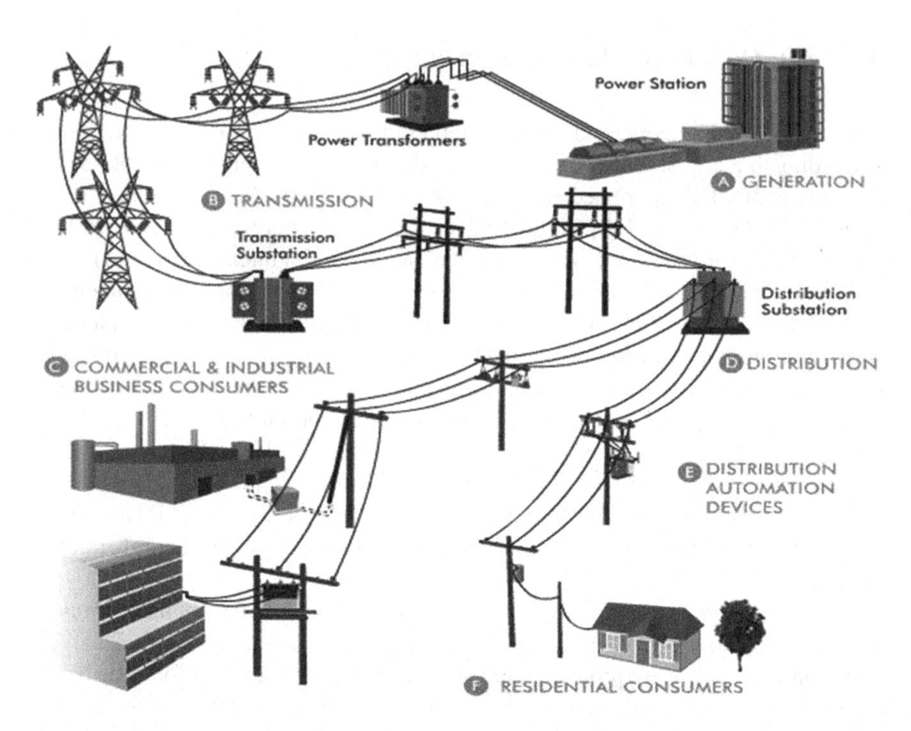

Fig. 3 Electrical grid utilizing big data analytics [30]

the power company to prioritize the work and maintain the most critical pieces of equipment first. This prioritization is vital given the limited number of human resources available to perform all of the required maintenance work. Big data and associated predictive analytics have the potential to reduce the cost of energy to customers and improve reliability of the electrical grid.

GE's (General Electric's) Predix is an example of a big data application in energy. According to GE, by connecting industrial equipment, analyzing data, and delivering real-time insights, Predix-based applications are unleashing new levels of performance of both GE and non-GE assets [31, 32]. As depicted in Fig. 4, data from a piece of wind turbine equipment can be captured by the Predix machine and sent to a cloud-based platform for storage. Applications can then access the data and perform analysis regarding the health and maintenance required on a piece of equipment. With Predix capturing data at all power stations (fossil, nuclear, wind, hydro, natural gas, solar, etc.), this big data provides the company with vast analytic capabilities across its entire fleet. Similar to the function provided by IBM's Watson Health for "human healthcare," electrical and other industrial companies may be able to use tools like Predix to proactively predict and address "equipment health" issues before the system is disrupted.

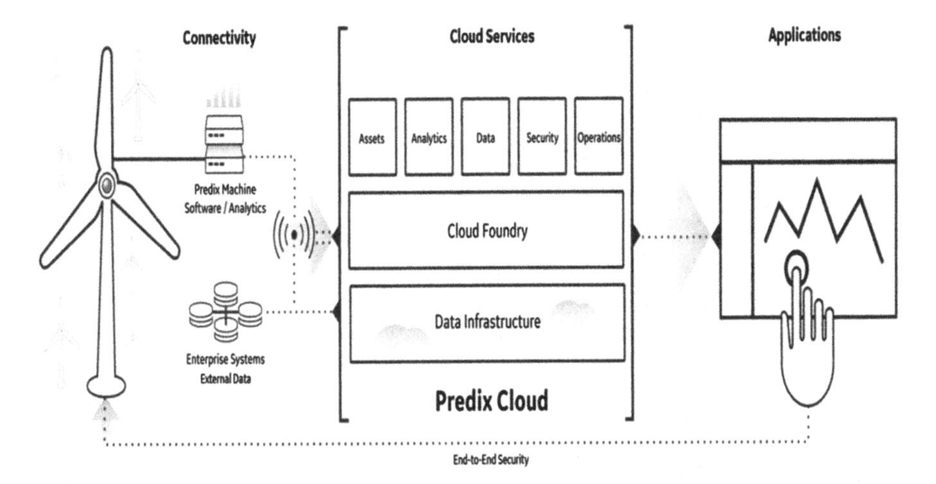

Fig. 4 GE Predix depiction [32]

3.3 Big Data in the Business Discipline

According to Harvard Business Review, businesses are collecting more data than they know what to do with [33]. To turn all this information into competitive gold, they need new data scientists to support their business flow. The additional roles such as data analytics professionals, database engineers, and data storage engineers must be integrated into the fabric of the company. In IBM's article, Gartner estimated that 4.4 Million data scientists were needed by 2015 and only 1/3 will be actually filled [34]. These new roles will enable the company to exploit the vast new flows of information and radically improve a company's performance. From a business standpoint, publicly traded companies comprise major portions of both the healthcare delivery and energy delivery systems. Big data is increasingly vital to the profitability of these companies since they must meet revenue growth projections and provide a return on investment to shareholders. As noted in the energy section, most electrical energy providers are also publicly traded businesses and are affected by big data. According to an EPRI (Electric Power Research Institute) report, the hype about big data often eclipses the basic tenants that underpin the utility business model—to provide electric services to customers and to earn a reasonable rate of return for shareholders [10]. This new data will probably not produce a huge profit but it is likely to increase the number of service offerings available to customers. Additionally, utilities (and other businesses) need help sorting out what is big "relevant" data and what is just a lot of unusable data.

As depicted by Fig. 5, businesses now have access to many sources of data including social media forms such as Facebook and Twitter, along with web site visits, and live streaming feedback, and all of this data available from anywhere in the world. Businesses must be able to correlate and effectively analyze this data

Fig. 5 Big data source variety [35]

to develop actionable intelligence. In turn, this actionable intelligence may lead to greater insight into the customer while providing an opportunity for improved profitability for the company.

Finally, as industries become more competitive and business models change, controlling costs becomes paramount in order to generate a reasonable rate of return. Deloitte identified five steps a company can utilize to become insight-driven – set objectives (define strategic priorities), identify the right data and tools, develop the approach (for acquiring and enriching data), go discover (reveal insights and test the data), and finally adopt the successful approaches [36].

4 Convergence: A Key Link Between Big Data, Platforms, and People

Convergence principles are a key link between big data, platforms, and people. Similar to big data, convergence has many definitions and continues to evolve. For instance, convergence is defined as a multi-disciplinary approach to problem solving that requires the fusion of technology, devices, knowledge, process and people from multiple engineering disciplines [9] and across data boundaries to solve societal challenges. Additionally, convergence is also defined as the integration of value considerations from multiple engineering disciplines to solve a common problem [9]. Yet another definition for convergence is the merging of distinct technologies, processing disciplines, or devices into a unified whole that creates a host of new pathways and opportunities [6]. The National Research Council (NRC) provides another definition for convergence as an approach to problem solving that cuts across disciplinary boundaries by integrating tools and knowledge to solve societal challenges [3]. While developed by different authors and organizations, all of the above definitions of convergence share key similarities that involve merging distinct technologies (tools, data, and devices) into a unified process while working across disciplinary boundaries. As the definitions indicate, solutions to today's

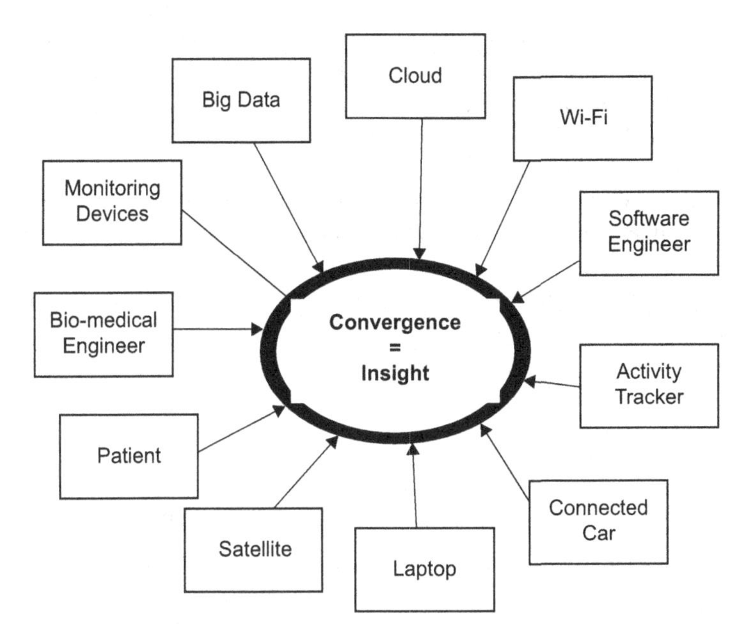

Fig. 6 Convergence leads to transdisciplinary insight

complex problems require engineers from multiple disciplines (chemical engineers, electrical engineers, biomedical engineers, and software engineers), big data and analytics professionals, along with business leaders, project managers, sociologists, and educators. Effective convergence demands the utilization of integrated systems with *big data storage* and the associated big data applications. These big data applications are vital to modern problem solving techniques which are increasingly convergent in nature. Therefore, as illustrated in Fig. 6, a thorough understanding of convergence principles allows solutions to healthcare, energy, and business problems to be developed with a better understanding of all the interdependent elements (discipline-specific knowledge, mobile devices, monitoring devices, big data repositories, customers, data transmission platforms, etc.) that add to the complexity and contextual diversity of the overall solution.

5 Conclusion

As described in this chapter, big data applications are critical to key industries such as healthcare, energy, and other businesses. This new generation of applications allows both structured and unstructured data to be collected from many disparate sources (devices, pieces of equipment, people, web sites, social media sites, and locations) and subsequently analyzed to make future predictions. We observed the historical evolution of information characterization as defined by data volume, velocity, variety, variability, and veracity. The ability to glean valuable insight and predict future outcomes using this vast amount and diversity of data in real-time has become invaluable in individual, yet interdependent disciplines with transdisciplinary applications now possible with convergence implications. These big data applications, as described in healthcare, engineering, and business, epitomize the key principle of convergence by allowing problem solving to occur across disciplinary, platform, and data boundaries. This treatise highlights the new opportunities that arise when big data advances occur in one discipline with spillover benefits from other disciplines when the properties of transdisciplinarity with respect to convergence are applied.

References

1. IBM: The four Vs of big data. Retrieved from: http://www.ibmbigdatahub.com/infographic/four-vs-big-data (2016)
2. SDPS: Society for Design and Process Science homesite. Retrieved from: www.sdpsnet.org (2017)
3. National Research Council: Convergence: facilitating transdisciplinary integration of life sciences, physical sciences, engineering, and beyond. National Academies, Washington, DC (2014), ISBN: 978-0-309-30151-0
4. Tanik, U.J., Ertas, A.: Healthcare engineering: implementing a transdisciplinary education and research program. In: Suh, S., et al. (eds.) Biomedical Engineering: Health Care Systems, Technology and Techniques. Springer Science + Business Media, LLC, New York (2011), DOI 10.1007/978-1-4614-0116-2_5
5. Fielder, D., Tanik, M., Tanik, U.J., Gattaz, C., Sobrinho, F.: Convergence Implications for Healthcare, Energy and Other Complex Systems. Society for Design and Process Science, Dallas (2016)
6. Massachusetts Institute of Technology: The convergence of life sciences, physical sciences, 406 and engineering. Retrieved from http://www.aplu.org/projects-and-initiatives/research-science-and-technology/hibar/resources/MITwhitepaper.pdf (2011)
7. Fielder, D., Garrett, D., Sobrinho, F.: Value-Based Process Engineering. Institute of Electrical and Electronics Engineers, New York (2015)
8. Fielder, D., Tanik, M., Tanik, U.J., Gattaz, C., Sobrinho, F.: Mobile healthcare delivery: a dynamic environment where healthcare, mobile technology, engineering, and individual lifestyles converge. In: SoutheastCon, Charlotte, NC, USA, 30 March–2 April 2017. Institute of Electrical and Electronics Engineers, Piscataway, NJ (2017). doi: https://doi.org/10.1109/SECON.2017.7925358
9. Fielder, D., Gattaz, C., Tanik, U.J., Tanik, M., Sobrinho, F.: Transdisciplinary Convergence – A Vital Consideration in Engineering Solutions. Institute of Electrical and Electronics Engineers, New York (2016)
10. Morris, H.D., Ellis, S., Feblowitz, J., Knickle, K., Torchia, M.: A software platform for operational technology innovation. Retrieved from https://www.ge.com/digital/sites/default/files/IDC_OT_Final_whitepaper_249120.pdf (2014)
11. SAS: Big data – What it is and why it matters. Retrieved from: http://www.sas.com/en_us/insights/big-data/what-is-big-data.html (2016)
12. Merriam-Webster: Big Data. Retrieved from: https://decisionviz.wordpress.com/2014/05/20/merriam-webster-dictionary-adds-big-data/ (2014)
13. Oracle: The foundation of data innovation. Retrieved from: http://www.oracle.com/big-data/index.html (2016)

14. Gartner: Retrieved from: https://research.gartner.com/definition-whatis-big-data?resId=3002918&srcId=1-8163325102 (2016)
15. CLP, C., Zhang, C.Y.: Data-intensive applications, challenges, techniques, and technologies: a survey on big data. Inf Sci. **275**, 314–347 (2014)
16. Kempe, S.: A short history of data warehousing. Retrieved from: http://www.dataversity.net/a-short-history-of-data-warehousing/ (2012)
17. Devlin, B.A., Murphy, P.T.: An architecture for a business information system. IBM Syst. J. Netw. Manag. **27**(1), 60 (1988)
18. Weilbarh, J., Viktor, H.: A data ware house for policy making – a case study. In: Proceedings of the 32nd Hawaii International Conference on System Sciences (1999)
19. Centers for Medicare & Medicaid Services: Retrieved from: https://www.cms.gov/research-statistics-data-and-systems/statistics-trends-and-reports/nationalhealthexpenddata/nationalhealthaccountshistorical.html (2016)
20. National Cancer Institute: Cancer prevalence and cost of care projections. Retrieved from: http://costprojections.cancer.gov/ (2016)
21. American Diabetes Association: The cost of diabetes. Retrieved from: http://www.diabetes.org/advocacy/news-events/cost-of-diabetes.html (2016)
22. American Heart Association: Heart disease and stroke statistics. Retrieved from: http://circ.ahajournals.org/content/early/2013/12/18/01.cir.0000441139.02102.80.full.pdf (2014)
23. Centers for Disease Control and Prevention: Chronic disease overview. Retrieved from: http://www.cdc.gov/chronicdisease/overview/ (2015)
24. U.S. Department of Health and Human Services: Aging statistics. Retrieved from: http://www.aoa.acl.gov/Aging_Statistics/index.aspx (2016)
25. Slidesharecdn: System overview of ubiquitous healthcare service. Retrieved from: http://image.slidesharecdn.com/healthcare-150616111313-lva1-app6892/95/toward-ubiquitous-healthcare-services-with-a-novel-efficient-cloud-platform-5-638.jpg?cb=1438614404 (2016)
26. Kevinmd: A new ecosystem of disruptive business models must arise. Retrieved from: http://cdn1.kevinmd.com/blog/wp-content/uploads/innovators-prescription-new-wave-of-disruptive-models-in-healthcare2.jpg (2016)
27. IBM Watson Health: IBM Watson Health. Retrieved from: http://www.ibm.com/watson/health/ (2016)
28. U.S. Energy Information Administration: Total energy annual energy review. Retrieved from: https://www.eia.gov/totalenergy/data/annual/showtext.php?t=ptb0105 (2012)
29. Power Technology: Power plant O&M: how does the industry stack up on cost? Retrieved from: http://www.power-technology.com/features/featurepower-plant-om-how-does-the-industry-stack-up-on-cost-4417756/ (2016)
30. Texas Electricity Alliance: Retrieved from: https://texaselectricityalliance.files.wordpress.com/2011/09/grid.jpg (2016)
31. GE Predix: Predix – industrial cloud. Retrieved from: https://www.ge.com/digital/predix (2016)
32. GE Predix Developer Network.: Retrieved from: https://www.predix.io/ (2016)
33. Harvard Business Review: Retrieved from: http://www.rosebt.com/uploads/8/1/8/1/8181762/big_data_the_management_revolution.pdf (2012)
34. IBM Big Data: What is big data? Retrieved from: https://www.ibm.com/big-data/us/en/ (2015)
35. Post control marketing: How Amazon uses big data to make you love them. Retrieved from: http://www.postcontrolmarketing.com/MediaSociety/2015/07/29/how-amazon-uses-big-data-to-make-you-love-them/ (2015)
36. Lewis, H.: Technology to accelerate discovery – investing in data analytics, pp. 1–29. Deloitte – The Institution of Engineering and Technology, Bareilly (2014)
37. Diebold, F.X.: A personal perspective on the origin(s) and development of 'big data': the phenomenon, the term, and the discipline, second version. PIER Working Paper No. 13-003 (2012). Available at SSRN: https://ssrn.com/abstract=2202843 or https://doi.org/10.2139/ssrn.2202843

A Big Data Analytics Approach in Medical Imaging Segmentation Using Deep Convolutional Neural Networks

Zheng Zhang, David Odaibo, Frank M. Skidmore, and Murat M. Tanik

Abstract Big data analytics uncovers hidden patterns, correlations and other insights by examining large amounts of data. Deep Learning can play a role in developing solutions from large datasets, and is an important tool in the big data analytics toolbox. Deep Learning has been recently employed to solve various problems in computer vision and demonstrated state-of-the-art performance on visual recognition tasks. In medical imaging, especially in brain tumor cancer diagnosis and treatment plan development, accurate and reliable brain tumor segmentation plays a critical role. In this chapter, we describe brain tumor segmentation using Deep Learning. We constructed a 6-layer Dense Convolutional Network, which connects each layer to every subsequent layer in a feed-forward fashion. This specific connectivity architecture ensures the maximum information flow between layers in the network and strengthens the feature propagation from layer to layer. We show how this arrangement increases the efficiency during training and the accuracy of the results. We have trained and evaluated our method based on the imaging data provided by the Multimodal Brain Tumor Image Segmentation Challenge (BRATS) 2017. The described method is able to segment the whole tumor (WT) region of the high-grade brain tumor gliomas using T1 Magnetic Resonance Images (MRI) and with excellent segmentation results.

1 Introduction

Deep Learning is a developing big data analytic method increasingly used in the field of computer vision. Deep Learning networks show excellent performance in many visual recognition tasks, such as in image classification, object detection and image segmentation. Convolutional Neural Networks (CNN) are a specific

Z. Zhang (✉) • D. Odaibo • M.M. Tanik
Department of Electrical and Computer Engineering, University of Alabama at Birmingham, Birmingham, AL, USA
e-mail: cheung@uab.edu; godaibo@uab.edu; mtanik@uab.edu

F.M. Skidmore
Department of Neurology, University of Alabama at Birmingham, Birmingham, AL, USA
e-mail: fskidmore@uabmc.edu

S.C. Suh, T. Anthony (eds.), *Big Data and Visual Analytics*,
https://doi.org/10.1007/978-3-319-63917-8_10

tool in the field of Deep Learning. Inspired in the late 1960s by the mammalian visual cortex, Hubel and Wiesel suggested a hierarchy of feature detectors in the visual cortex [1]. The receptive field is a particular region of the sensory space in which a stimulus will fire that neuron. In CNN, the computationally generated "artificial neurons" are designed to be very sensitive to a given "receptive field." Not surprisingly given the inspiration, CNNs increasingly have proved to be exceptional computational tools for visual image segmentation and classification tasks. In the past three years, Convolutional Neural Network models have achieved state-of-the-art performance in various computer vision tasks. For example, Karen last name and Andrew investigated the relationship of depth and accuracy of neural networks and proposed a VGG neural network architecture which won the classification and localization in the ImageNet Challenge in 2014 [2, 3]. Alex, Ilya and Geoffrey developed a seven hidden layers neural network architecture and trained it to classify 1.3 million high resolution images into 1000 different classes [4]. Christian Szegedy proposed a novel inception module in its 22 layer GoogleNet architecture which has achieved a 6.67% error rate in training 1.2 million images into 1000 classes [5]. Recently, Olaf Ronneberger proposed a special CNN architecture called a "U-Net." The U-Net contains what is termed a "contracting" path and an "expanding" path. Unlike other CNNs, which can lose spatial localization information, this specific architecture enables the information flowing between the contracting path and the expanding path which retains spatial localization [6]. More recently, Gao et al. introduced a dense Convolutional Neural Network called DenseNet, which obtained significant improvements over the state-of-the-art computer vision classification tasks. This architecture is proposed based on the observation that the results of CNNs become more accurate and the networks become more efficient to train if they contain shorter input and output connections between layers [7].

We show in this chapter how these methods can be applied to a medical imaging task (segmentation of brain tumor). This is a pertinent and current task, as to date there are multiple semi-automatic or automatic brain tumor segmentation methods, with various success rates [8]. We approach this field by evaluating segmentation of brain gliomas. Gliomas typically manifest in adults and can be measured by MRI with multiple sequences, such as T2-weighted fluid attenuated inversion recovery (Flair), T1-weighted (T1), T1-weighted contrast-enhanced (T1Gd), and T2-weighted (T2), as illustrated in Fig. 1. Tumor glioma segmentation is challenging, as the appearance of gliomas is similar to gliosis and stroke in MRI data [9]. Additionally, gliomas have varied shapes and sizes, can occur throughout the brain, and frequently have undefined boundaries related to invasion of surrounding brain tissues.

Dvorak et al. attempted to solve a multi-class brain tumor segmentation task by training CNNs using local label patches [10]. Very deep CNN models were proposed and implemented by Pereira et al. [11]. Darvin et al. proposed a framework of 3D CNN models for brain tumor segmentation [12]. Most of these segmentation methods used image patches as inputs when training the CNNs. One of the disadvantages is these methods did not consider the tumor's appearance and spatial consistency and assumed that each voxel's label is independent.

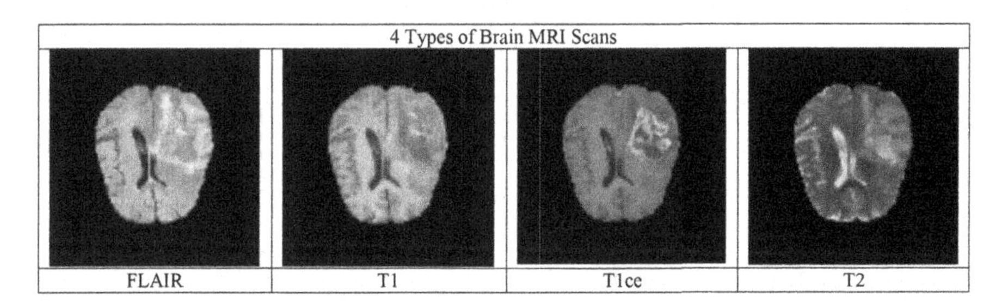

Fig. 1 Fluid attenuated inversion recovery (FLAIR), T1-weighted contrast-enhanced (T1ce), T2-weighted (T2), T1-weighted (T1) MRI images

Inspired by the success of Deep Learning techniques in computer vision and brain tumor segmentation, we proposed and implemented a 6-layer Dense Convolutional Network [7] to perform the brain tumor segmentation task using the MICCAI BraTS 2017 dataset [13]. Our contribution to the community can be summarized as following:

1. We implemented a 6-layer dense convolutional network to perform brain tumor segmentation task, which has not yet been reported in lecture.
2. We used the full MRI images instead of image patches as input data, which preserves the information of tumor appearance and its spatial consistency.
3. We segmented the whole tumor (WT) region using our model with excellent results.

2 Methods and Materials

2.1 Imaging Data

All the brain tumor MRI data used in this study were obtained from the MICCAI 2017 Multimodal Brain Tumor Segmentation Challenge (BraTS2017) [13], which also includes the previous imaging data from both the MICCAI BraTS 2012 [14] and MICCAI BraTS 2013 [15]. The data contains real volumes of 210 high-grade glioma patients and 75 low-grade glioma patients [16]. These images are in a commonly used "nifty" format, however in this study, we converted images into ".png" format. Each MRI image volume contains 155 axial slices (see Fig. 1 for an example of axial brain images). There are four types of MRI scans for each patient: (1) T2-weighted fluid attenuated inversion recovery (Flair), (2) T1-weighted (T1), (3) T1-weighted contrast-enhanced (T1ce), and (4) T2-weighted (T2). Figure 1 demonstrates these four types of MRI scans. All images were skull stripped, and we had access to manually generated segmented "ground truth" labels, produced by expert board-certified neuropathologists [17]. The label included four types of tumor regions: (1) edema, (2) non-enhancing solid core, (3) necrotic core, and (4) enhancing core.

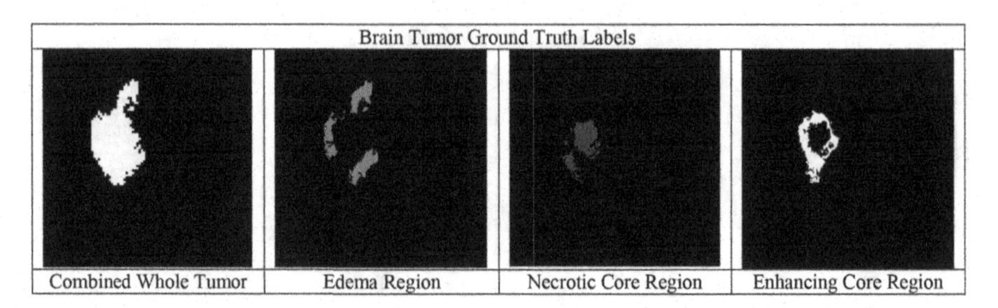

Fig. 2 Example of ground truth label and combined whole tumor

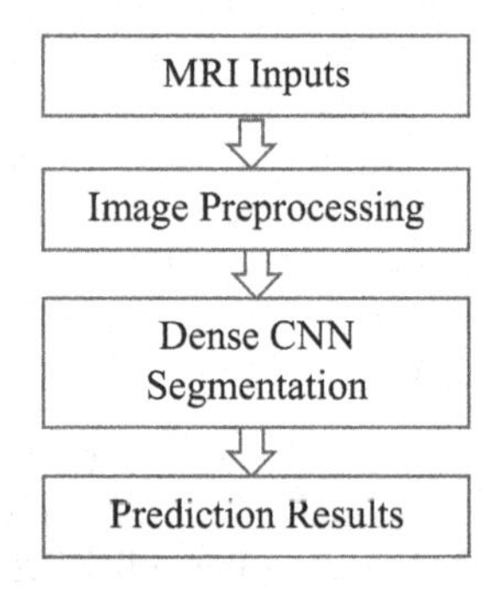

Fig. 3 Flowchart of the proposed process

For demonstration purposes, we combined these four tumor region labels to create a "Whole Tumor" (WT) label. Figure 2 illustrates the three types of the tumor regions and a combined WT region. We organized the training dataset based on existence of WT tumor in each MRI slice. The training dataset consisted of 14,076 labeled high-grade T1-weighted (T1) MRI slice images with tumor regions. The test dataset consists of 200 high-grade T1-weighted MRI slice images with tumor regions.

2.2 *Brain Tumor Segmentation Method Based on Dense Convolutional Neural Network*

Our brain segmentation method consists of four main steps. First, inputs of MRI image were normalized in order to balance the bias of the image. Second, we expanded the training dataset by applying a data augmentation method. Third, we built a 6-layer dense convolutional neural network and trained our model by feeding single MRI slice data and its corresponding whole tumor ground truth labels. Fourth, we predicted segmented results were generated with the test dataset. Figure 3 illustrates flow chart of the proposed process.

2.3 *Image Preprocessing*

One of the challenges in working with MRI data is its varying intensity. Variations in intensity are produced either by patient movement or environmental noise (such as radiofrequency noise), which can have a great effect on image intensity. Normalization is a process that changes the range of pixel intensity values. After converting from "nifti" format to ".png" format, MRI slices have a size of 240 by 240 pixels, and are single channel grayscale images. The mean of all the 14,076 images is 30.133432 and the standard deviation is 58.69165. We normalize the image by first using the image pixels minus the mean value, then divided by the standard deviation.

Another challenge in training neural network is overfitting, which means that the prediction results fit too closely or exactly to a particular dataset and fail to predict additional data reliably.[1] Data augmentation is a way to prevent overfitting, by using the information of existing training data to create more training data. In our experiment, we applied cropping, flipping and rotating to our training dataset. Figure 4 demonstrates our data augmentation outputs.

2.3.1 Dense Convolutional Neural Network Model

Once we have completed pre-processing, we applied our CNN. The proposed Deep Learning model is a 6-layers dense convolutional neural network model. As shown in Fig. 5, all layers are connected with each other in a feed-forward fashion. $x_0, x_1, \ldots, x_5$ represent the inputs from the layers of $0, 1, \ldots, l-1$. For example, the input of x_3 includes the output from layer 0, input from layer 1, input from layer 2 and output from layer 2. We concatenated these layers in our implementation. This special connectivity pattern improves the information flow, including location information, between layers. In traditional feed forward convolutional neural networks architecture where the information is carried from

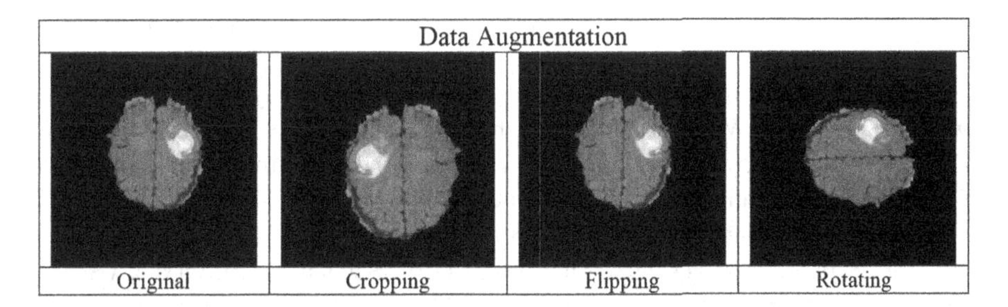

Fig. 4 Outputs of the processed imaging modalities

[1]Definition of "overfitting" at OxfordDictionaries.com: this definition is specifically for Statistics.

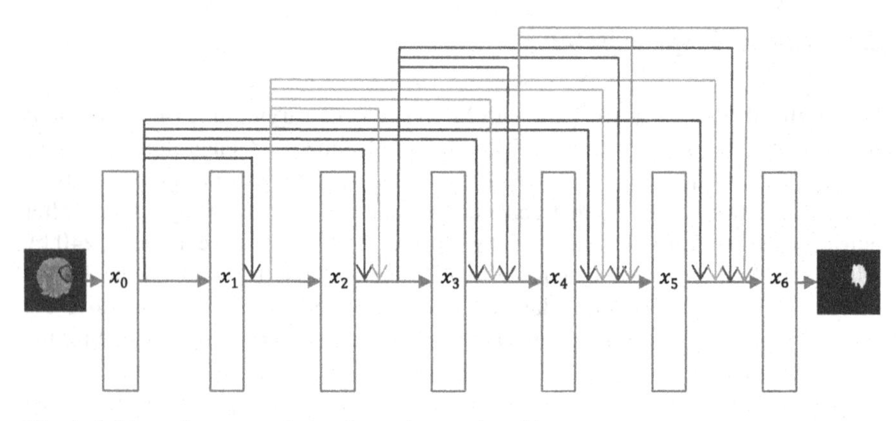

Fig. 5 A 6-layer dense convolutional neural network architecture

Table 1 The architecture of 6-layer dense convolutional neural network

Layers	Inputs	Outputs
Inputs	Size (1, 256, 256)	N/A
Conv1	Inputs	Conv1
BatchNorm	Conv1	Conv1
Conv2	Conv1	Conv2
Conv3	Conv2	Conv3
BatchNorm	Conv3	Conv3
Dense2	Conv2, Conv3	Conv4
Conv4	Dense2	Conv4
BatchNorm	Conv4	Conv4
Dense3	Conv2, Conv3, Conv4	Dense3
Conv5	Dense3	Conv5
BatchNorm	Conv5	Conv5
Dense4	Conv2, Conv3, Conv4, Conv5	Dense4
Conv6	Dense	Conv6

layer to layer, the neural work must learn redundant feature maps from the previous layer. Dense convolutional neural networks require much fewer parameters because of its connectivity between layers [7].

The input of our network 256 by 256. For convolutional layers, the kernel size we have is 3×3. We also applied batch normalization after each convolutional layer. After each convolutional layer, we concatenate the multiple inputs to create a single tensor input and feed into the next convolutional layer (Table 1). Our model is implemented using Keras [18]. The computing unit we used to train the neural network is p2.8xlarge NVIDIA Titan X GPU on Amazon Web Service (AWS) instance. The runtime environment is Linux Ubuntu 14.04. The performance of our model is measured by the dice score [19].

$$Dice\,Score\ (A, B) = \frac{2 \times both\ (A, B)}{only\,A + only\,B + 2 \times both\ (A, B)}$$

The dice score compares the size of the overlap of the two segmentations divided by the total size of the two objects. A refers to the "ground truth". B refers to the segmented prediction. Both (A, B) refer to the total number pixels which have the value 1 in both A and B.

3 Experiment and Results

We show that the proposed architecture, applied to the BRATS 2017 dataset, demonstrates high quality tumor segmentation (Fig. 6) The first column is the actual patient data. The second column is the ground truth label of the combined Whole Tumor (WT). The third column is the segmented results generated by our dense convolutional neural network model. We trained our model with three epochs, each epoch takes about 4 h.

We evaluate the performance by calculating the dice score and the accuracy between our segmented results and the ground truth label. We obtain the average dice Score of 0.7223 and accuracy 0.9841 for the whole tumor segmentation among 200 test MRI images.

4 Conclusion

In this study, we proposed a novel Deep Learning model by constructing a 6-layers dense convolutional neural networks to perform brain tumor segmentation tasks using the MICCAI BraTS 2017 dataset. We have demonstrated that our proposed model is able to perform brain tumor segmentation task and has remarkable results.

Compared with of existing CNNs based brain tumor segmentation methods, most are using image patches as inputs. They first classify the image patches into different tumor classes, such as necrosis, edema and health tissue, then label the center voxel of the image patches according to these predicted classes assuming that each voxel label is independent to perform the tumor segmentation. However, the tumor regions are not independent. In our model, we train the CNNs using the full size MRI images as input, which will take the considerations of the brain tumor's spatial consistency and its appearances information. Secondly, the special connectivity pattern of our network greatly improves the information flow between layers in the network, which makes the network easy to train. The layers are directly connected with each other. One big advantages of this architecture is that it requires fewer parameters than traditional feed forward CNNs. Our segmented results have achieved the average dice coefficient score of 0.7223 and accuracy score of 0.9841. Currently, our model only takes single MRI slice as input data, in future work, we plan to include the information between slices and improve the segmentation results.

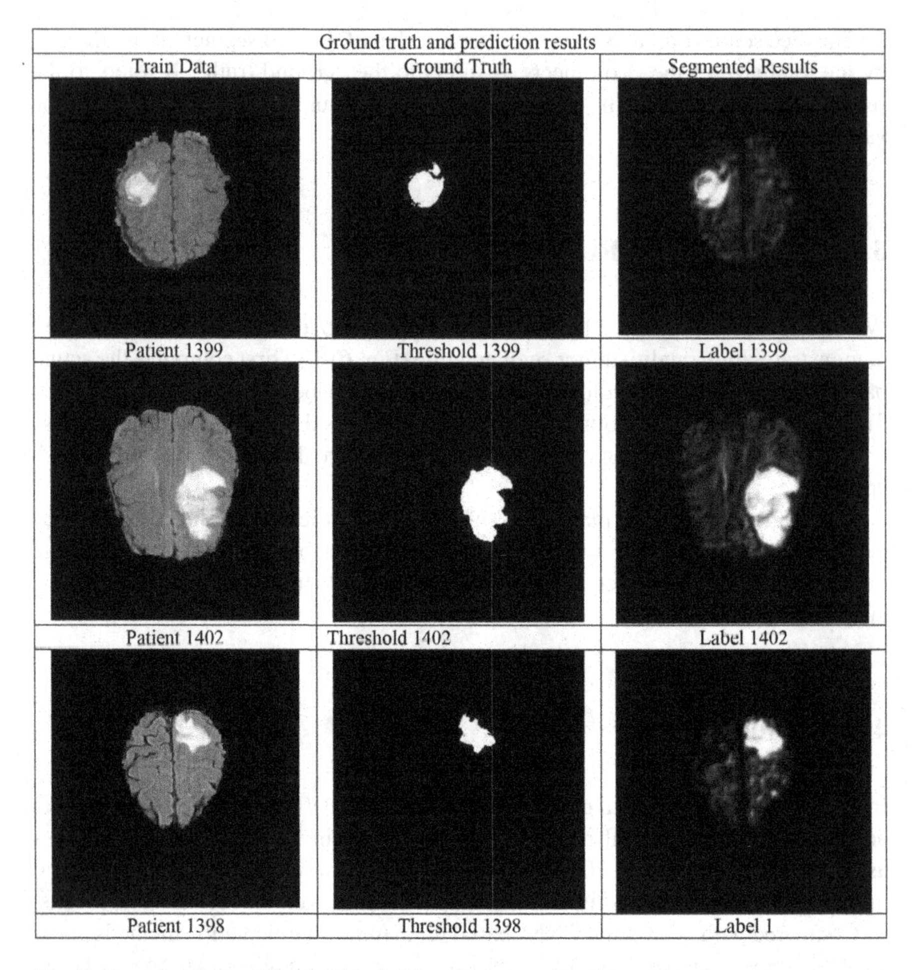

Ground truth and prediction results
Train Data
Ground Truth
Segmented Results
Patient 1399
Threshold 1399
Label 1399
Patient 1402
Threshold 1402
Label 1402
Patient 1398
Threshold 1398
Label 1

Fig. 6 Example of comparison of threshold prediction mask and ground truth

References

1. Hubel, D.H., Wiesel, T.N.: Receptive fields of single neurones in the cat's striate cortex. J. Physiol. **148**(3), 574–591 (1959)
2. Simonyan, K., Zisserman, A.: Very deep convolutional networks for large-scale image recognition. arXiv preprint arXiv:1409.1556 (2014)
3. Simonyan, K., Zisserman, A.: Very deep convolutional networks for large-scale image recognition. arXiv:1409.1556 (2015)
4. Krizhevsky, A., Sutskever, I., Hinton, G.E.: Imagenet classification with deep convolutional neural networks. In: NIPS'12 Proceedings of the 25th International Conference on Neural Information Processing Systems, vol. 1, pp. 1097–1105 (2012)
5. Szegedy, C., et al.: Going deeper with convolutions. In: Proceedings of the IEEE Conference on Computer Vision and Pattern Recognition, pp. 1–9 (2015)

6. Çiçek, Ö., et al.: 3D U-Net: learning dense volumetric segmentation from sparse annotation. In: Ourselin, S., Joskowicz, L., Sabuncu, M., Unal, G., Wells, W. (eds.) International Conference on Medical Image Computing and Computer-Assisted Intervention Lecture Notes in Computer Science, vol. 9901. Springer, Cham (2016)
7. Huang, G., et al.: Densely connected convolutional networks. arXiv preprint arXiv:1608.06993 (2016)
8. Ronneberger, O., Fischer, P., Brox, T.: U-net: convolutional networks for biomedical image segmentation. In: International Conference on Medical Image Computing and Computer-Assisted Intervention, pp. 234–241. Springer, Cham (2015)
9. Goetz, M., et al.: DALSA: domain adaptation for supervised learning from sparsely annotated MR images. IEEE Trans. Med. Imaging. **35**(1), 184–196 (2016)
10. Dvorak, P., Menze, B.H.: Local structure prediction with convolutional neural networks for multimodal brain tumor segmentation. In: Menze, B., et al. (eds.) Medical Computer Vision: Algorithms for Big Data. MCV 2015 Lecture Notes in Computer Science, vol. 9601. Springer, Cham (2016)
11. Pereira, S., et al.: Deep convolutional neural networks for the segmentation of gliomas in multi-sequence MRI. In: International Workshop on Brainlesion: Glioma, Multiple Sclerosis, Stroke and Traumatic Brain Injuries. Springer, Cham (2015)
12. Darvin, Y., et al.: 3-D convolutional neural networks for glioblastoma segmentation. arXiv preprint arXiv:1611.04534 (2016)
13. MICCAI: BraTS 2017 dataset (July 25, 2017). http://braintumorsegmentation.org/ (2017)
14. MICCAI: BraTS 2012 dataset (July 25, 2017). http://www2.imm.dtu.dk/projects/BRATS2012/ (2017)
15. MICCAI: BraTS 2013 dataset (July 25, 2017). http://martinos.org/qtim/miccai2013/ (2017)
16. Havaei, M., et al.: A convolutional neural network approach to brain tumor segmentation. In: International Workshop on Brainlesion: Glioma, Multiple Sclerosis, Stroke and Traumatic Brain Injuries. Springer, Cham (2015)
17. Shin, H.-C., et al.: Deep convolutional neural networks for computer-aided detection: CNN architectures, dataset characteristics and transfer learning. IEEE Trans. Med. Imaging. **35**(5), 1285–1298 (2016)
18. Keras: The Python Deep Learning Library: (July 25, 2017): https://keras.io/ (2017)
19. Dice, L.R.: Measures of the amount of ecologic association between species. Ecology. **26**(3), 297–302 (1945)

Big Data in Libraries

Robert Olendorf and Yan Wang

Abstract The term Big Data is somewhat loose. Roughly defined, it refers to any data that exceeds the users ability to analyze it in one of three dimensions (the three Vs): Volume, Velocity and Variety. Laney [1, 2] Each of these has different challenges. Huge volumes of data require the ability to store and retrieve the data efficiently. High velocity data requires the ability to ingest the data as it is created, essentially very fast internet connections. Highly variable data can be difficult to organize and process due to its unpredictability and unstructured nature. Bieraugel [3] Also, multiple data streams can be combined to answer a variety of question. All forms of big data can require high performance computing and specialized software to analyze. Given the fuzziness of defining big data,

1 Introduction

The term Big Data is somewhat loose. Roughly defined, it refers to any data that exceeds the users ability to analyze it in one of three dimensions (the three Vs): Volume, Velocity and Variety. Laney [1, 2] Each of these has different challenges. Huge volumes of data require the ability to store and retrieve the data efficiently. High velocity data requires the ability to ingest the data as it is created, essentially very fast internet connections. Highly variable data can be difficult to organize and process due to its unpredictability and unstructured nature. Bieraugel [3] Also, multiple data streams can be combined to answer a variety of question. All forms of big data can require high performance computing and specialized software to analyze. Given the fuzziness of defining big data,

Libraries can have to primary interests in big data. First is using big data to help with their operations. Libraries can use data to optimize their collections, better utilize space, asses their instruction and to provide information to their users. As more data is recorded in libraries, and more data is available externally via both

R. Olendorf
PennState University, Pennsylvania, PA, USA

Y. Wang (✉)
University of Alabama at Birmingham, Birmingham, AL, USA
e-mail: yanwang3@uab.edu

S.C. Suh, T. Anthony (eds.), *Big Data and Visual Analytics*,
https://doi.org/10.1007/978-3-319-63917-8_11

open and commercial sources, the opportunity for using these data streams is both a promising avenue for libraries, but also potentially risky in terms of allocating resources.

The second area in which libraries can work with big data is in the field of research data services. More and more funding agencies require data from funded research to be made public. Researchers often find themselves without sufficient skills and resources to adequately manage the data during their projects, and even more commonly do not have the skills, time or resources to prepare the data for archiving and often also are unable to find appropriate repositories for their data. Libraries have always played the role of housing societies academic and research output. While in previous generations, this has primarily focused on books and articles, expanding the scope to research data is a natural extension to the libraries role.

Here we explore these to realms of big data use in libraries. We explore the benefits, costs and risks associated with using big data. We also provide use cases, guidance to getting started and briefly outline tools and resources for working with big data. McCoy et al. [4] Given the speed with which the technology can change, we focus more on general guidance. Where specific technologies are referenced, it should be understood that technological progress can render some of the discussion obsolete, however, the general concepts should still be applicable.

2 Using Big Data In Libraries

Libraries are increasingly interested in using data to better manage their collections, utilize space and assess their services. Chen et al. [5] Traditional library services alone can generate huge amounts of data, especially for larger libraries. Some of the data are obvious, such as circulation data, patron counts and knowledge database usage. Other data streams might not be so obvious, such as harvesting social media feeds. It can be difficult for libraries to fully utilize the data create, and including additional data can present further difficulties.

However, using data to improve and optimize library function is a real benefit to be considered. Before diving into using data, it is important for the library to understand the full extent of the issue. Libraries should understand what data streams they wish to include. They should be aware of the costs in terms of both staffing and infrastructure will be required. They should also have a good idea of the benefits to be realized.

Big data projects can be of two types, a one-off project, or a standing service. The development of a either will generally take the following steps the biggest difference is a standing service will require a more durable production data store while a one off project can use less resource intensive data stores. One-off projects are ideal for answering questions that only need to be answered once. However, if the need is to repeat analyses, or have data ready for a variety of analyses or real time analyses, then a standing service should be preferred.

Either type of project will typically follow these steps should follow the order outlined. For instance, if benefits aren't identified early on, there is a risk of spending a great deal of time on resources that has no real need.

1. Identify benefits
2. Identify current resources
3. Identify solutions
4. Identify unmet needs
5. Implementation

2.1 Needs Analysis

Using Big Data can sound sexy and current. The temptation is often to assume that there are benefits because other people are doing it, it's in the news or because its the new technology. It is important to first understand the potential benefits and current needs. This will not only prevent unnecessary allocation of resources, but it will also result in a better product.

Ideally, a comprehensive list of current or potential services should be drawn up, with associated problems, difficulties these services face due to lack of sufficient information or analysis. In addition, ideas on how services can be improved with additional data should be considered as well. While not all of the listed items will directly involve big data, it is likely many will. Also, the list shouldn't, at this point, consider costs or even feasibility. The idea is to get a good list to work with.

Anytime a service can be improved or created through the application of data and analysis is an opportunity to use Big Data.

2.2 Current Resources

A comprehensive list of resources should be developed as well. Again, be inclusive here. Include any resource that could potentially be relevant. This will be useful both in estimating the cost of a solution and identifying solutions that fit well with resources you already have.

Computing resources should also be cataloged. Servers, storage and software should all be noted. Include cloud based resources that you currently use. List any database software (MySQL, PostGres), operating systems, statistical and analysis software you own or are licensed to use.

Staffing resources are critical. Big Data and data analysis are new to many libraries. However, there may be untapped skills within your library. Also, you should ensure that you list all of the Information Technology staff you currently employ. Stanton [6]

The first type of resource to consider is the data itself. A list of current data streams the library currently creates or has access to. This might include current holdings, circulation statistics, gate counts, patron counts, reference interactions and online chat. Also include external data sources which you may currently subscribe to as well as social media feeds.

A final critical resource to consider are external resources. Perhaps the most important would be the High Performance Computing (HPC) center and Cyber-Infrastructure (CI) center at your institution. Additional resources might be Statistical consultants and Risk Management.

2.3 *Identify Solutions*

With your needs in hand, you can start planning solutions. The solution will typically involve several components.

- Data store
- Administration
- Policies and access
- Analysis and visualization
- Query access (optional)

These may seem pretty standard and they are. However, when working with big data care must be taken to choose the right solutions. As noted previously the solutions you choose will depend on the nature of the project, one-off or service based. Ideally you would tackle your data store first.

2.4 *Data Stores*

For one-off solutions, the type of data store may be less critical. CSV or excel files may be sufficient for some projects, but they may not scale well with true big data projects, even one-offs, so it may be necessary to use SQL databases or document based databases such as Mongo DB or Hadoop. The cost of making a mistake isn't so great however, mainly in extra time spent fixing the mistake or time spent working with an inefficiency.

Service level data stores must be reliable and robust. Unfortunately, they also typically require maintenance and development. Also, because the the data store is intended for long term use, significant time in planning and development will be required. The benefit, of course, is the ability to make use of the data for a variety of purposes continually over time. Service level big data stores, often utilize technologies such a data warehouse architecture or document based or NoSQL databases Hecht and Jablonski [7].

Data Warehouses take data from a variety of sources, transform it and then load it into a central data store Inmon [8]. The focus with a Data Warehouse is to transform or pre-process the data so that it is more efficient for the types of queries you will perform. For instance, data is often timestamped. Libraries might be interested in how a variety of resources are used on an hourly basis over the course of a week. With just a time stamp this can be an expensive query, especially if you are merging different data streams, where you must first determine the day of the week and hour of the day for each record before joining the records together. Using Data Warehouse type architecture, the data is preprocessed, for instance assigning each record a day of the week and an hour based on its timestamp.

Document based repositories (NoSQL) forgo the structure of data base schema and are often said to scale better than SQL. Additionally, products such as Kibana for Elasticsearch are designed for big data analysis and visualization.

The choice will depend on a variety of factors. Before deciding it is important to determine your needs and also the technical skills and support your institution can provide for a given solution.

2.5 Administration

For service level data stores you will need to provide an administrative layer. Some solutions come with built in administrative front ends. Ideally, when researching your data store, you should also determine what administrative tools are available. Not having to build your own can save a great deal of time and resources.

2.6 Policies and Access

One important non-technical issue to resolve are policies of use and access. At the start, a document that states the scope of the project, and policies governing use should be drafted. The scope specifies the data that are expected to be included in the data store and what they will be used for. Defining the scope helps to focus development and limit" scope creep" which happens as new ideas come to the light. Stating the scope also helps better set stakeholder expectations. Policies concerning use will help define the users, and also provide avenues for interested parties to gain access to the data.

One very important aspect to consider is how to handle sensitive data. This includes data with personally identifying information, trade secrets or other data that could be potentially damaging to other agents.

When mashing different data sets together, additional care must be taken that the combined data doesn't cause additional risks. For instance, data that identifies a student's school may not be risky. Data about student's ethnicity or economic status may not be risky. Combining the two however, may suddenly make it much easier to identify individuals with a fair degree of certainty.

2.7 Analysis and Visualization

The types of analyses and visualizations that are needed may drive the solutions chosen. While most solutions provide a wide array of analyses and solutions some are limited in what they can do.

Also, the skills of the potential users should be taken account. Analysis and Visualizations can be divided into two types of interfaces, point and click, and scripted. Point and click interfaces are easier to use, especially for users with little or no background in data analysis. The trade-off is that these are usually more limited in what they will do, and perhaps more importantly they often make it difficult document how the analyses and visualizations were created. Also, the ease of entry can allow new users to use the analyses and visualizations wrongly.

Scripted solutions essentially require the user to write short programs to accomplish the analyses and visualizations. They require considerable more skill to use. However, thcy typically allow the user to a perform a much wider range of analyses and visualizations. Because the analyses are written as a script, it is easy to know how they are done and additional documentation can usually be added as needed. Because the user will typically already be fairly skilled in analysis and visualization, the risk of inappropriate analyses is reduced. Using scripted solutions should be considered a best practice, but may not be practical for all institutions.

Analysis solutions should also be evaluated on how the results can be exported. Some solutions only allow tables to be exported in CSV or even proprietary formats. Graphs and charts may only be exportable in PNG or other image formats. If one of the goals is to make data available on the web and to make it interactive, solutions should be chose that provide the ability to export content as interactive graphs usually in SVG.

2.8 Query Access

Query access is providing refers to providing a low level interface to the data, usually in the form of an HTTP based API. This feature is not necessary, but many data warehouses provide. Implementing this requires some thought, including access control.

Query access can also be used to allow the extraction of data to be used in online visualizations. For instance, a good API can generate json outputs that can then be input into a product such as Highcharts that generates very nice and interactive charts and graphs.

2.9 *Implementation*

One-off research projects are often practical, and several examples of research projects conducted from within libraries exist [citations needed]. Typically, these projects do not require extensive technical infrastructure, and if the required expertise does not exist within the library, collaborations with other departments or libraries will typically suffice.

Service level projects, will exceed the capabilities of many libraries to support them with the necessary infrastructure and expertise. However, collaborations with other institutional departments such as high performance computing, advanced cyber-infrastructure, or similar units may be advantageous.

3 Library Big Data Services

Libraries are in a unique position to not only use big data for their own purposes, but also to provide services around big data. Many libraries are already offering research data services such as data management planning, data collection, data curation and data archiving. Sugimoto et al. [9] Some libraries also offer some level of analytic and visualization support, much of this driven by demand created by various mandates to provide data management plans and make data more accessible. At the same time, researchers are constantly pushing the limits of their data sets, increasing both their size and the complexity. While in some realms, this has been going on for some time, other domains are just starting to explore the possibilities of big data. Libraries can play an important role in facilitating their institutions research, although as before, this will often require cooperating with other units.

We will explore some of the possible services libraries can provide using the data life cycle developed by DataONE 1. While libraries can play an important role, libraries will not have great a role in each stage of the life cycle. Some stages may present too great of challenges or require too many researchers, so libraries will need to choose carefully what services to provide and how to resource them.

3.1 *Planning*

Assisting with Data Management Plans (DMPs) is one of the most common services offered by academic libraries. Working with a researcher on a big data project only differs from typical DMPs in several ways. Depending on the funder, a typical DMP covers most of the topics Most of the details will be covered below. Librarians who assist in writing DMPs involving big data projects will need to be aware of all the services the researcher might require to adequately address the issues in a DMP. Also, they should be aware that much of the boiler plate text that many libraries

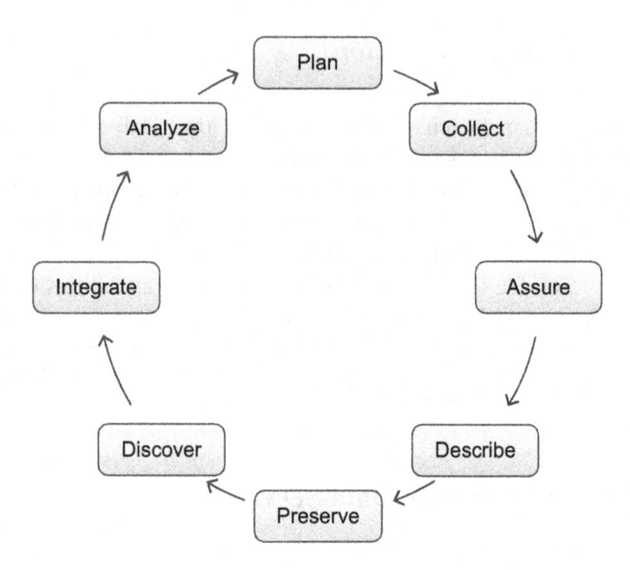

Fig. 1 The data life cycle developed by DataONE

provide may not be applicable. For instance, many libraries provide text describing how their institutional repository can accept data, however, big data projects may be too large or complex for these repositories (Fig. 1).

3.2 *Collection*

Collection of big data may seem to be outside the range of typical library services. However, there are some opportunities to be found. Some libraries may already have in house skills common big data sources, such as social media feeds. Libraries some commercially available big data sources might be acquired and treated as part of the library collection.

Acquiring and working with big data can be computer intensive. Libraries can collaborate with other units such as high performance computing to provide computing assistance and also the technical infrastructure to work with big data. Libraries primary role here might be to ensure that the data collected is properly organized and documented to ensure that it can be easily archived later.

3.3 *Assure, Analyze and Integrate*

Quality insurance (QA), analysis and integration of the data is typically the responsibility of the researcher. However, if librarians are supplying data as part of their collection, there may be some responsibility to perform quality assurance on this data and provide adequate documentation for researchers to conduct their own QA methods. Also, in general, libraries should provide a supporting role by providing tools and resources to aid researchers in conducting and documenting their QA protocols. This might include version control, tutorials, and workshops.

3.4 Describe

Providing data with adequate metadata and documentation fall squarely within the strengths of libraries and this is a task researchers struggle with providing proper documentation and metadata as well. Libraries should provide services to help researchers document their big data and supply it with metadata. Here we differentiate between documentation and metadata. Documentation is human readable information that provides details about various aspects of the data. Metadata is typically machine readable information, often machine generated and in addition to providing some information that aids in using the data, also functions to identify and describe the data.

Ideally, librarians would work proactively with researchers so that they begin documenting their work as it is created. Otherwise, especially with very large complex projects, much information is lost or forgotten over the course of the project. Also, if researchers wait until the end of the project to document their work, it can become overwhelming and often does not get done. Two fairly simple steps to take at the start of the project are to create a README text file, and to create an initial file structure and naming convention. Big data projects can be large and complex and often involve multiple collaborators. By starting the project in an organized fashion with solid documentation, the project will typically proceed far more smoothly than otherwise.

README files are simple text files frequently used in software development to describe the project and document how to use the software. The contents can vary quite a bit, but the overall goal of the README is to provide enough information that other researchers can replicate, validate and reuse the data with confidence. The README should be updated throughout the project and acts as both a means to communicate with the current collaborators and also as a method for providing needed information to others who may use the data in the future. Some common components include

- A list of creators with roles contact information.
- A list of contributors.
- A description of the project, similar to an article abstract.
- A description of the data that includes.
- A description of the contents.
- A description of the naming conventions.
- How to view the data, what software is needed.
- How to run the analysis to replicate the data.
- License information
- An example citation

The project's file structure and naming conventions should also be set at the beginning of the project. A simple, high level directory structure should be created with naming conventions that help to describe the contents and purpose of the directory, such as "data", "analyses", "output". Likewise, naming conventions for other directories and their files should be consistent and provide information about what they are.

3.5 Preserve

Along with data description, archiving and preservation is perhaps the other primary service libraries can provide. Many funders, including most STEM agencies such as NSF and NIH require that researchers make their data available upon completion of the project. NIH [10] For big data projects this can be especially problematic for the researchers due to the quantity and complexity of the data. It can also present unique problems to the libraries.

Many libraries have institutional repositories and some also have dedicated data repositories. However, in most cases they are limited by their technology to smaller and simpler datasets. This presents a problem to researchers who are using big data and whose data sets may not easily fit into the institutional repository solutions. An additional difficulty occurs when data is derived from commercial sources such as financial data or social media feeds. In these cases, researchers are limited by what they can make public, while still often required to release enough information to validate their results.

To better handle big data sets in repositories there are several possible paths. First, it may often be possible to reduce the size of the data by reducing repetitive content, or removing content that can be generated again. For instance, Twitter data can be reduced to a list of just the Tweet IDs vastly reducing the amount of storage needed for the data. This is also another opportunity to collaborate with other institutional units such as who have more resources in the form of storage and computing power. Often these units lack the skills in curation to make the data findable and reusable which libraries often have. Finally, there may be other external repositories that are more appropriate for the data. This is common in the genomics realm for instance.

3.6 Discover

Aside from some of the more common big data sources like Twitter, discovering useful data can be difficult. Librarians involved in research data support can serve a useful role in working with researchers to find relevant and useful datasets. These can range from support of genomics research to locating corpora for digital humanities projects. In addition to locating the datasets, librarians as informaticists should also be familiar with techniques for efficiently acquiring the data, storing and processing the data.

Another issue with data discovery is negotiating intellectual property issues. While much data in the STEM disciplines may be open, there is considerable data that comes with a variety of licensing arrangements. Librarians can be helpful to researchers in understanding these, but importantly, also to determine how to best make the resulting data as open as possible so that the research can be disseminated in as open a manner as possible.

4 Conclusion

Although it has not always been recognized as such, data has always been fundamental to academic research. While we typically think of data as numbers and measurements, it also includes text, images and anything else that is used to inform our understanding of the natural world, and the artifacts we create. As the digital age moves on, we are increasingly seeing data exposed as part of the process and the understanding that access to the data increases our ability to understand what has been done and also to advance faster by reusing data that has already been created.

In the last 15 years or so, we have seen an increase in the velocity, volume and variety of data created to the point where our ability to comprehend, let alone work with the data stretches our limits. Librarians have always been custodians of our shared store of knowledge. As the digital and data age move forward, librarians must adapt to not only handling information in the form of written text but also working with information in all its variety and volume. This will require partnerships with other experts such as high performance computing, cyber-infrastructure and also the researchers themselves.

References

1. Laney, D.: META delta. Application delivery strategies, 949:4. (2001)
2. McCoy, C., Marcinkowski, M., Sawyer, S., Sanfilippo, M.R., Meyer, E.T., Rosenbaum, H.: Social informatics of data norms. Proc. Assoc. Inform. Sci. Technol. **53**(1), 1–4 (2016)
3. Bieraugel, M.: Keeping Up with Big Data. (2013)
4. Rosenbaum, H.: Social informatics of data norms. Proc. Assoc. Inform. Sci. Technol. **53**(1), 1–4 (2016)
5. Chen, H.-I., Doty, P., Mollman, C., Niu, X., Yu, J.-C., Zhang, T.: Library assessment and data analytics in the big data era: practice and policies. Proc. Assoc. Inform. Sci. Technol. **52**(1), 1–4 (2015)
6. Stanton, J.M.: Data science: what's in It for the new librarian? https://ischool.syr.edu/infospace/2012/07/16/data-science-whats-in-it-for-the-new-librarian/ (2012). Accessed 12 Jan 2017
7. Hecht, R., Jablonski, S.: NoSQL evaluation: a use case oriented survey. In: Proceedings—2011 International Conference on Cloud and Service Computing, CSC 2011, pp. 336–341 (2011)
8. Inmon, W.: Building the Data Warehouse. Wiley, Hoboken, NJ (2005)

9. Sugimoto, C.R., Ding, Y., Thelwall, M.: Library and Information Science in the Big Data Era: Funding, Projects, and Future [A Panel Proposal]. (2012)
10. National Institutes of Health: Nih Data Sharing Policy. https://grants.nih.gov/grants/policy/data_sharing/ (2006). Accessed 14 Jan 2017

A Framework for Social Network Sentiment Analysis Using Big Data Analytics

Bharat Sri Harsha Karpurapu and Leon Jololian

Abstract Traditionally, surveys were used as one of the major methods for finding out the opinion of a group of people about a particular topic. However, over the last two decades with the proliferation of Web to the social media sites such as Twitter, Facebook, and Tumblr, social media are increasingly becoming the platform of choice for people to express their views or opinions. With an account of over two billion users, social media provides a major source for gathering people moods and opinions. Several public and private organizations, such as Government and companies, are attempting to exploit the expressed preferences, opinions, and attitudes regarding politics, commercial products and other matters of personal importance for a competitive edge. One of the efficient ways to get this information is by performing sentiment analysis on these electronic repositories. With the data being ubiquitous, the bottlenecks here are processing speed, storage, and time, which are involved with the traditional storage system. So to deal with the data processing of these massive amounts of data, some special tools and techniques are offered by Big Data framework.

1 Introduction

Traditionally, surveys were used as one of the major methods for finding out the opinion of a group of people about a particular topic. However, over the last two decades with the proliferation of Web to the social media sites such as Twitter, Facebook, and Tumblr, social media are increasingly becoming the platform of choice for people to express their views or opinions. With an account of over two billion users, social media provides a major source for gathering people moods and opinions. Several public and private organizations, such as Government and companies, are attempting to exploit the expressed preferences, opinions, and attitudes regarding politics, commercial products and other matters of personal importance for a competitive edge. One of the efficient ways to get this information

B.S.H. Karpurapu (✉) • L. Jololian
The University of Alabama at Birmingham, Birmingham, AL, USA
e-mail: harsha99@uab.edu

S.C. Suh, T. Anthony (eds.), *Big Data and Visual Analytics*,
https://doi.org/10.1007/978-3-319-63917-8_12

is by performing sentiment analysis on these electronic repositories. With the data being ubiquitous, the bottlenecks here are processing speed, storage, and time, which are involved with the traditional storage system. So to deal with the data processing of these massive amounts of data, some special tools and techniques are offered by Big Data framework.

This article is based on my master thesis which describes the process of developing a generalized big data framework for performing social network sentiment analysis as a POC (Proof of Concept). This framework consists of a real-time collection of data followed by machine learning methods for performing sentiment analysis on a big data platform. For getting data and for performing sentiment analysis, a live mini-blogging website 'Twitter' is considered which generates almost 6000 tweets per second with a restriction to express their opinion in 140 words. For processing speeds and storage capacity, large-scale cluster-computing framework like the spark was used, for its efficiency to handle data and for its speed. For sentiment analysis on the collected tweets, Naive Bayes algorithm was used. At the end of the article, we presented with the results of the two case studies to demonstrate the effectiveness of this framework.

2 Research Questions of Thesis

2.1 *Question 1: How Would It Be Possible to Parameterize Domain Characteristics in Order to Allow the Framework to Specialize for Various Domains?*

The whole idea of this thesis is based on this question, and the answer to this is "Yes." The framework, which we developed here can be used for performing sentiment analysis on various domains such as Banking, Automobiles and Airlines by changing domain characteristic values. Here in this thesis, we performed sentiment analysis on Automobile Industry as a POC (Proof of Concept) and as case study I. This framework was extended at the end with an action email system, which we performed on Regions Bank as a case study II.

2.2 *Question 2: How Would It Be Possible to Create an Upgradable Framework Such that Components Can Be Replaced Whenever New Tools with Better Performance Become Available?*

Yes, we can build an upgradable framework. Here in the development of this framework, we used different tools such as Spark, Jupyter Notebook, Anaconda, Tableau, and Plotly. However, for developing the framework, we followed a layered architecture, which makes it possible to replace any layer or any part in the layer with a new efficient tool/technology in the future.

3 Domain Literature and Theoretical Background

3.1 Sentiment Analysis

Sentiment Analysis is one of the fields in the area of natural language processing which aims at determining the writer's attitudes, emotions or moods from the text regarding various topics. One of the best known measures of sentiment is the polarity. The simplest form of the polarity is to have only two degrees either positive or negative: either like or dislike button in Facebook, we can consider them as the extreme ends of the continuous scale. The continuous scale can have some discrete points, which result in rating scales such as Amazon rating scale and IMDB rating scale. Generally, semtiment analysis is performed by extracting the subjective information from the given text and people opt for different methods for this process, for example: Agarwal, in his paper [1] extracted the subjective information from text by POS tagging for performing sentiment analysis.

3.1.1 Types of Sentiment Analysis

- Document Level Sentiment Analysis, deals with determining whether the whole document is expressing positive, negative or neutral opinion about an entity. The main goal is to extract the overall polarity of the document [2].
- Sentence Level Sentiment Analysis, deals with the analysis of determining whether the particular sentence is expressing a positive, negative or neutral opinion. This can also be considered as a standard text classification problem.
- Aspect Level Sentiment Analysis, deals with the type of analysis which aims at determining the the sentiment of individual sentences and phrases about a particular aspect. This is also called as a fine grained sentiment analysis [3].

3.2 Machine Learning

Arther Samuel, the pioneer in the field of Artificial Intelligence, defined Machine Learning as "A field of study that gives the computers ability to learn without being explicitly programmed". In a generalized sense, one can understand Machine Learning as the process of making a machine to learn from the given situations/instances through different algorithms and enabling it to perform on its own in the future through the acquired knowledge/intelligence.

3.2.1 Learning Process

A typical machine learning process involves training from the given input data, developing a hypothesis model and predicts the future variable. The set of all the equations that we fit the training data is referred to as the hypothesis space. The one best-fitted equation which reduces the risk or cost on the overall training data, is taken as the Final Hypothesis model.

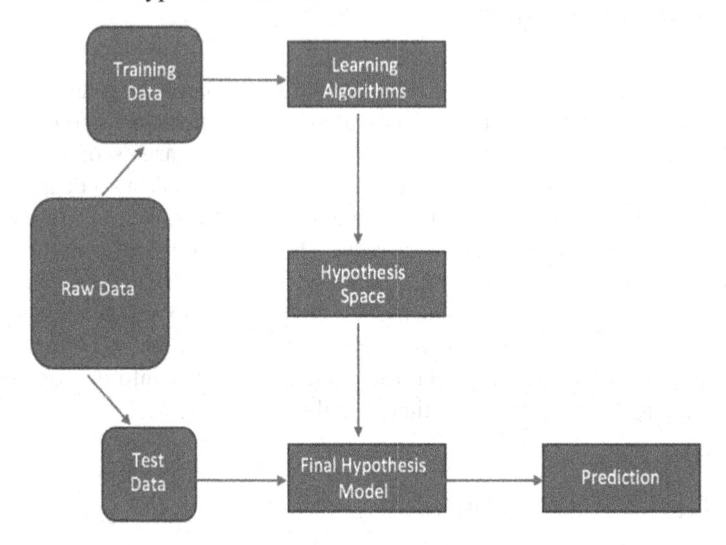

The complete Loss or Cost or Risk Function is given by

$$R_D\left(h_{i(x)}\right) = \frac{1}{N}\sum (y_i - h_1(x_i))^2$$

where R_D is the Risk function over the Data 'D' having 'N' data points.

y_i is the observed value (target or label) and $h_1(x_i)$ is the predicted value. Now the final hypothesis 'g' is the one, who is having lowest risk function in the whole hypothesis space 'H' (represents all the polynomial spaces)

$$g(x) = argmin_{h(x)\in H} R_D\left(h(x)\right)$$

so g(x) is the final hypothesis model which best fits the given data. During the training process, various considerations are to be taken for bias and variance just to make sure that the curve is not overfitting or underfitting the data.

3.2.2 Naive Bayes Algorithm

Naive Bayes algorithms is used for classification problems, which are claimed to be full probabilistic models based on Bayes theorem.

Bayes Theorem

Posterior probability of 'A' is equals to the prior probability of 'A' times the likelihood of 'A' divided by the evidence.

$$P\left(\frac{A}{B}\right) = \frac{P\left(\frac{B}{A}\right)P(A)}{P(B)}$$

here $P\left(\frac{B}{A}\right)$ represent the likelihood of 'B' given 'A', lets represent that as L(A). Then the above equation can be written as

Suppose A = Y (Parameter of data)

B = X (data) then the above proportionality changes to

$$P\left(\frac{B}{A}\right) \propto \mathrm{L(Y)}*P(Y)$$

i.e. the posterior distribution of the data is equal to the likelihood of the data for the given parameter times the prior probability of the parameter.

Spam Filter Equations by Naive Bayes

Assume that there are five words frequently appear in the spam mail and suppose that someone got a new mail which uses three words of the spam category and rest are not, then the probability of the given mail being spam is given by

$$P\left(\frac{spam}{W_1, W_2^c, W_3, W_4^c, W_5}\right) = \frac{P\left(\frac{W_1, W_2^c, W_3, W_4^c, W_5}{spam}\right)P(spam)}{P\left(W_1, W_2^c, W_3, W_4^c, W_5\right)}$$

Now by Naive Bayes assumption, let's consider the conditional independence of words in the spam and conditional independence of words in the non-spam. Then we can write the likelihood as

$$P\left(\frac{W_1, W_2^c, W_3, W_4^c, W_5}{spam}\right) = P\left(\frac{W_1}{spam}\right)*P\left(\frac{W_2^c}{spam}\right)\ldots..P\left(\frac{W_5}{spam}\right)$$

This might be a huge assumption, but it makes the computation much simpler. Similarly, the not spam likelihood is given by the below formula:

$$P\left(\frac{W_1, W_2^c, W_3, W_4^c, W_5}{not\ spam}\right) = P\left(\frac{W_1}{not\ spam}\right)*P\left(\frac{W_2^c}{not\ spam}\right)\ldots..P\left(\frac{W_5}{not\ spam}\right)$$

3.3 *Big Data Analytics*

Big Data refers to the huge volumes of Datasets, which cannot be handled by a single machine or any normal traditional database system. Gartner, the world's leading IT advisory and research company, defines Big Data as "Big data is high volume, high velocity, and/or high variety information assets that require new forms of processing to enable enhanced decision making, insight discovery, and process optimization." Some of the notable researches inluclude suresh [4]: developed a framework for analyzing consumer opinions in map reduce environments, and Amir [5]: proposed a approach for dealing storage constraint of these datasets by developing online analytical algorithms.

4 System Architecture

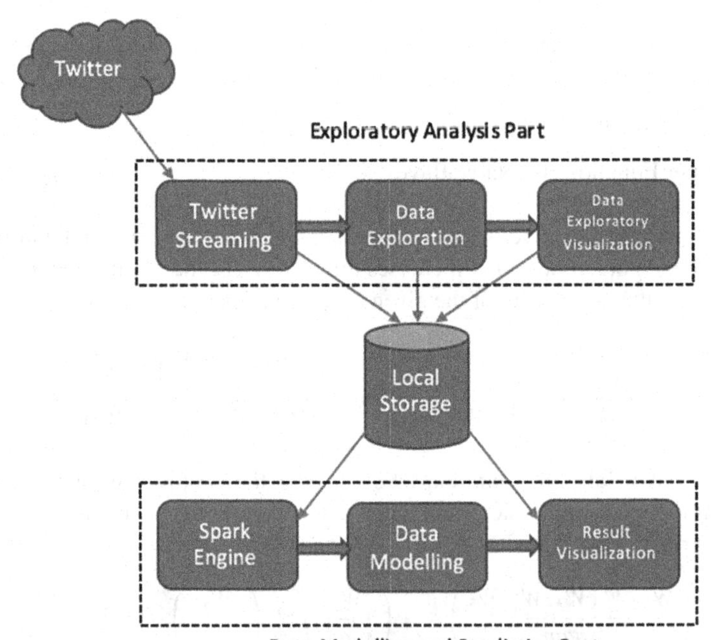

The architectural block diagram of the proposed framework will consist of a data source, which is a social networking site (Twitter), Exploratory Analysis Section, Modelling Section, and local storage.

The above figure represents the architectural block diagram of the proposed framework with all the components for input, processing, and output. Twitter Streaming is used to collect tweets from the Twitter, and the raw data of the collected tweets is stored in the local database. Data Exploration section is the preprocessing and analysis section, where the collected data is subjected to various cleaning and preparation steps to make it more human readable and ready for modeling part. The final cleaned output is sent to Visualization section, which shows some exploratory analysis with interactive visualizations using Tableau. Now the cleaned data is collected from the local storage of our system and sent to the spark engine to perform the predictive sentiment analysis. The final Output of the sentiment is visualized using Tableau plots. The outputs of every stage are stored in the local database for efficiency and documentation purposes.

5 Process Flow

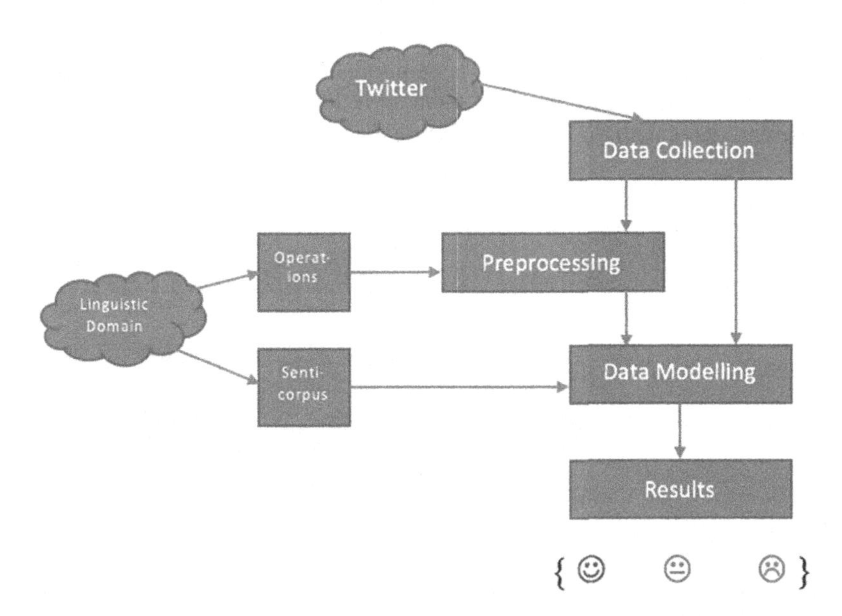

5.1 *Data Collection*

The input data source here is the tweets, and Twitter is the data repository. For data collection purposes, Twitter provides two API's, streaming API and REST API. Streaming API is used to collect tweets in real-time, whereas REST API provides the way for collecting tweets from the past week. Streaming API offers longer connectivity and provides data in almost real-time. Twitter search API is one of the REST API's of Twitter which almost equal to Twitter search functionality. The returned data is in the form of JSON or XML using GET requests. In most cases, the returned format will be in JSON, which is having more compactness.

At the end, the proposed framework is being implemented with two use cases. In the first use case, Twitter streaming API was used and in the second use case, Twitter search API was used. In both cases, we can pass a specific word on which the tweets have to be collected. In Twitter streaming, the keyword is called filter, and in search API it is called query word.

5.2 Preprocessing

5.2.1 Language Detection

The tweets collected from Twitter streaming will contain tweets with different languages from various countries. Therefore, the first step in our process should be to separate English language tweets from the collected Twitter corpus.

5.2.2 Lower Case

One way to make processes easier is to normalize the tweets. In this case, the tweets are normalized by converting them all to lower case which makes it easier for comparing the with the sentiment dictionary.

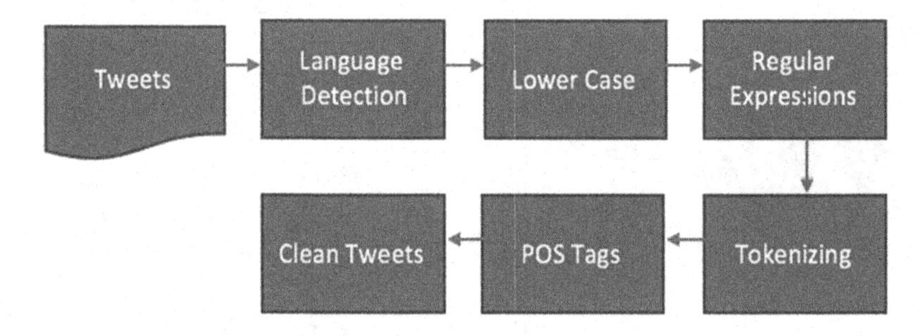

5.2.3 Regular Expressions

Regular Expressions are used to select a particular type of content according to the pattern given in the expression. In this case, regular expressions are used to remove punctuation, special character (except space).

5.2.4 Tokenization

Tokenizing is the process of breaking a sentence by the white spaces or any kind of special symbols. Each meaning word or symbol is called a token. Here tokenization was done by white spaces.

5.2.5 Parts of Speech Tagging

POS tagging is the process of assigning parts of speech tag to every word in the sentence. We used POS tagging to detect the subjective words from the tweets like the like adjectives, adverbs and verbs.

5.3 *Data Modelling*

Once tweets are cleaned the next step is the data modeling part, i.e. the sentiment assignment part. Here for sentiment polarity assignment, Naive Bayes theorem was used.

Initially, a word dictionary is maintained with all the positive and negative sentiment words with their log probabilities. When a new tweet is given the individual words are identified in negative and positive log probabilities dictionaries, and final negative log probability and positive log probability is calculated which is the sum of their individual log probability values. Final polarity is assigned according to the winning polarity log probability.

Suppose there are five subjective words$\{w_1, w_2, w_3, w_4, w_5\}$ in a tweet.

Total Negative Log Probability =
Neg_Log_Prob(w_1) + Neg_Log_Prob(w_2) + Neg_Log_Prob(w_3) +
Neg_Log_Prob(w_4) +Neg_Log_Prob(w_5)

Total Positive Log Probability =
Pos_Log_Prob(w_1) + Pos_Log_Prob(w_2) + Pos_Log_Prob(w_3) +
Pos_Log_Prob(w_4) +Pos_Log_Prob(w_5)

Final tweets polarity is positive if, total positive log probability is greater than total negative log probability else the final polarity is negative. If the none of the words are there in the word dictionary, then the final result is neutral.

6 Tools Profile

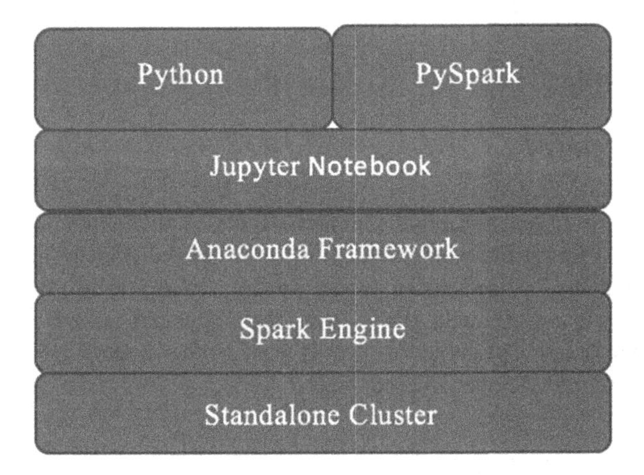

6.1 Jupyter Notebook

"Jupyter Notebook" can be considered as a web application or interactive computing environment that allows to create and share documents that contain live code, equations, visualizations and explanatory part. It consists of three important parts Notebook Web Applications, Kernels and Notebook documents.

6.2 Anaconda Framework

Anaconda is the product of Continuum Analytics, they define anaconda as "A free, easy-to-install package manager, environment manager, Python distribution, and collection of over 720 open source packages with free community support.

6.3 Spark

Spark is a general-purpose large-scale cluster-computing framework with many data tools stacked under a single umbrella for handling different cases. One of the most important aspects of spark is its scalability and efficiency. It is well built with a higher level API's which are available in different programming languages like Java, python, and scala. Data operations in spark are performed by transformations and actions. Unlike most frameworks, spark can disseminate these operations across various worker nodes by providing programming abstraction and speed by parallel processing.

7 Case Studies

7.1 Case Study I

In this case study, the proposed framework was implemented for doing the sentiment analysis for the whole automobile industry domain. In this part, tweets were collected in real time using twitter streaming on a total of nine different automobile companies (Honda, BMW, Kia, Audi, Toyota, Lexus, Hyundai, Buick, and Mercedes Benz). The subjective words from all the tweets were extracted and final polarity score was given to each tweet.

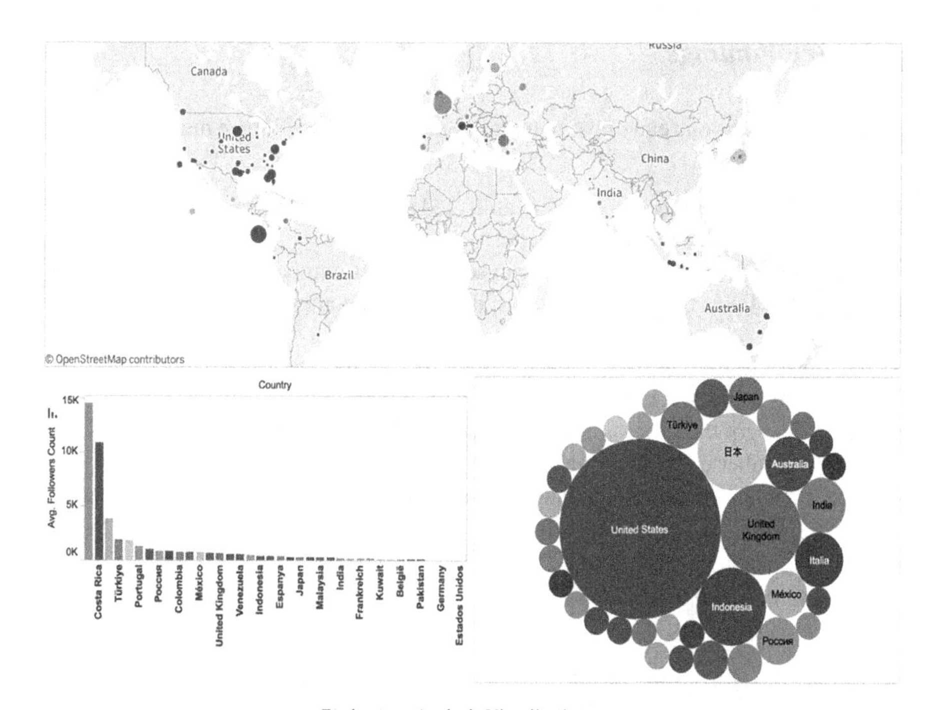

Exploratory Analysis Visualization

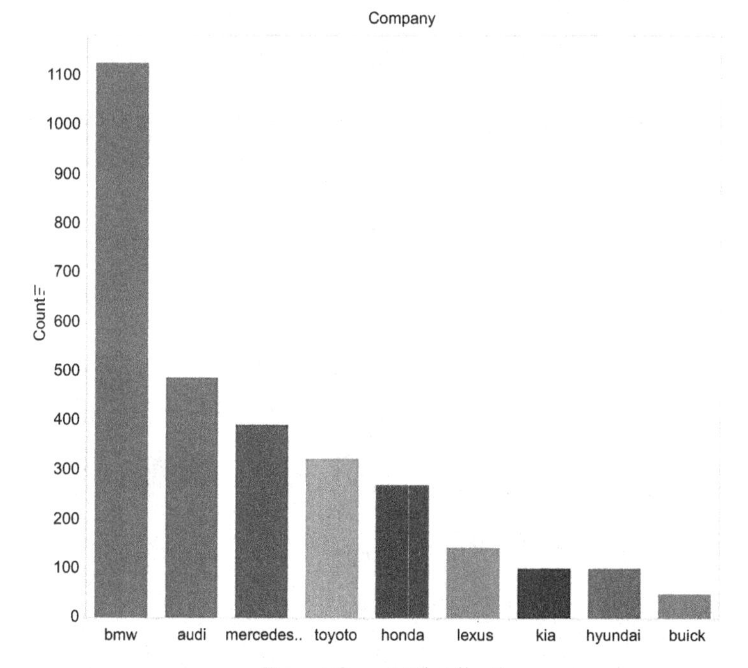

Tweet Count Distribution

7.2 Key Findings

From there graphs one can draw many inferences, However, We are mentioning only a few here.

- A total of 38 language tweets were collected and 3132 were English tweets which are almost 65% of the data collected.
- In Demographic distribution, we can see that most numbers of tweets were from the US. So US is the largest source of data when it comes to tweets of automobile industry.
- The average number of followers is more for the country Royaume-Uni, So sentiment expressed by these tweets is very important.
- BMW has got the more number of tweets with 1126 count, while Buick got the lowest number of tweets with 49 count. Therefore, we can say that BMW is the most popular one and Buick is the least popular one of all.

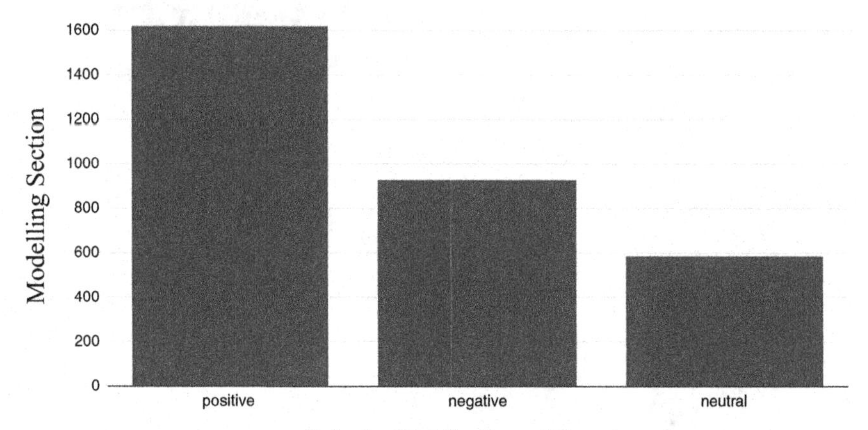

Polarity Distribution of Tweets

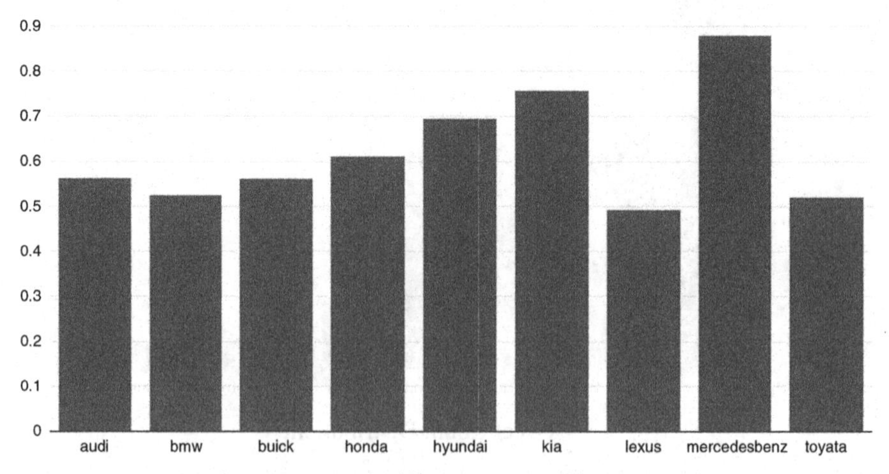

Sentiment Score of Different Automobile Companies

	text	company	tweet_words	polarity
0	RT @GuyKilfoil: The magnificent #bmw M550i XDr...	bmw	rt magnificent xdrive individual metallic s gu...	positive
1	https://t.co/AYIsnMdfeG #Carbon Fiber 2012+ #H...	honda	https tcoaylsnmdfeg carbon fiber + honda civic...	negative
2	Driving home on I-405. #bmw https://t.co/KEW8P...	bmw	home i https tcokewpvine	negative
3	Driving home on I-405. #bmw https://t.co/wErf7...	bmw	home i https tcowerfzbikr	negative
4	RT @bmwcanada: Radically more. The new #BMW #M...	bmw	new rt bmwcanada bmw m https tcosczpevxqo radi...	positive
5	RT @BMW_SA: The #BMW Concept X2 fuses a flat c...	bmw	flat robust rt bmw_sa bmw concept x coup silho...	positive
6	RT @theJDMculture: S T A N C E \n#is300\n#Lexu...	lexus	rt n lexus thejdmculture s c e theculture http...	neutral
7	RT @SAMAATV: The #BMW car that was gifted by #...	bmw	lakh rt samaatv bmw car hassannawaz maryamnawa...	negative
8	So much fun at the dc auto show #dcautoshow201...	audi	fun dc s auto dcautoshow moment appreciate aud...	positive
9	RT @theJDMculture: S T A N C E \n#is300\n#Lexu...	lexus	rt n lexus thejdmculture s c e theculture http...	neutral
10	Is it calling me? 😩 #Toyota #GT86 #topgearph ...	toyota	toyota soho central private gt topgearph resid...	positive
11	RT @antonioperic: Did you know that #bmw cars ...	bmw	php rt antonioperic bmw cars https tcoauaopccs...	negative
12	RT @leanettef: My babies are enjoying their 1s...	lexus	rt leanettef babies st flapanthers game flapan...	positive
13	Getting there...\n#cb750 #sohc #honda #caferac...	honda	cb sohc honda caferacer flashfirecoatings http...	neutral
14	RT @sapien_jorge: I really love Audi 😎😍#R8 #...	audi	rt sapien_jorge i audi r audi audi el paso htt...	positive
15	#lexus locklear pornstar most popular free por...	lexus	lexus popular free locklear pornstar porn http...	positive
16	BMW has some great Accessories. Check them out...	bmw	great accessories https tcoctbtckk bmw bmwseri...	positive
17	#Honda 1976 Honda CB HONDA CB360-T https://t.c...	honda	tcomecsthrriq honda honda cb honda cbt https r...	positive
18	Der neue BMW M760Li in Palm Springs - cooles A...	bmw	mli new mli der neue bmw palm springs auto coo...	positive
19	RT @leanettef: Period 1 of the @FlaPanthers ga...	lexus	selfie lexus rt leanettef period flapanthers g...	negative

Sample Output of Case Study I

Of all the automobile companies chosen, Mercedes Benz has the highest sentiment score of 0.8766 and Lexus has the lowest sentiment score of 0.492. Here, we can observe that, although BMW is popular one, the final sentiment score is highest for Mercedes Benz. So popularity is different from sentiment.

7.3 *Case Study II*

In this case study, the proposed framework was implemented for doing the sentiment analysis for a single entity. This part used Twitter search API, and extracted all the past tweets on a single entity Regions Bank. The proposed framework was extended in this case so that it can send action emails to the assigned persons depending on the sentiment threshold.

```
 (u'@askRegions that day and they also deposited 2.50 idk if it was to my account or someones else. It was on 2/10 i
 believe and i cant report\n',
  'neg',
  1.0),
 (u'@askRegions hey someone used my card at krogers and spent 1200$ my account info had to be stolen because i was tr
aveling to Atlanta all day\n',
  'neg',
```

Sample Output of Case Study II

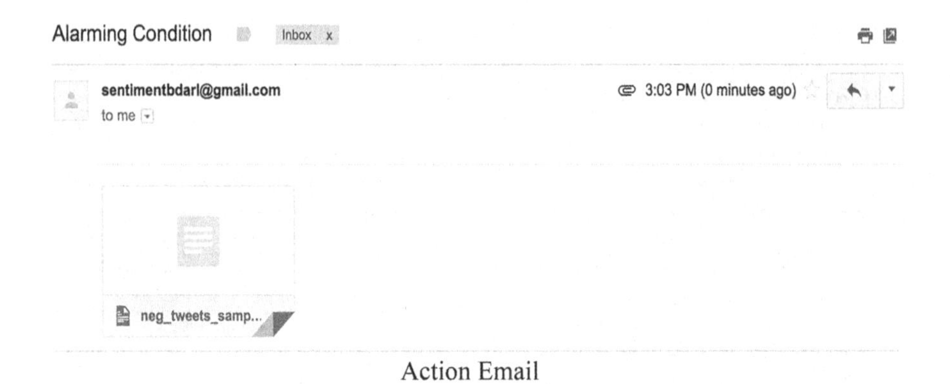

Action Email

8 Conclusions and Future Directions

This whole article is concentrated on developing a generalized framework for social network sentiment analysis. The main aim of this framework is to make it adaptable for performing the sentiment analysis on any domain such as an automobile, banking, and telecom. This generalized framework can be used by anyone as it mitigates the underlying technical aspect of the analysis process by providing an easy to use interface. Clearly, the effectiveness of this framework on the two case studies revealed its usability to other cases.

As of now, this framework is developed for Twitter, we can develop the framework to be suitable for other social networking sites such as Facebook and Instagram. We can also make it more appealing by creating a web console at the front end and code at the back end, where results will have automatically presented to users on the web console. Coming to the technical perspective, for sentiment analysis, we used unigram model. However, we can use bigrams or trigrams, which guarantee more accuracy in many cases. Currently, this framework was built to support only English language, Nevertheless, we can improve the framework to support multi-languages [6]. Although, we hard coded some visualization plots, we can make these things automated by purchasing some third party libraries.

References

1. Agarwal, A., Xie, B., Vovsha, I., Rambow, O., Passonneau, R.: Sentiment Analysis of Twitter Data. Springer, Heidelberg (2011)
2. Pang, B., Lee, L., Vaithyanathan, S.: Thumbs up?: sentiment classification using machine learning techniques. In: Proceedings of the ACL-02 Conference on Empirical Methods in Natural Language Processing, vol. 10, pp. 79–86. Association for Computational Linguistics, Stroudsburg, PA (2002)
3. Yang, B., Cardie, C.: Joint inference for fine-grained opinion extraction. In: ACL (1). pp. 1640–1649 (2013)

4. Suresh Ramanujam, R., Nancyamala, R., Nivedha, J., Kokila, J.: Sentiment analysis using big data. In: Computation of Power, Energy Information and Commuincation (ICCPEIC) (2015)
5. Hossein, A., Rahnama, A.: Distributed real-time sentiment analysis for big data social streams. In: Control, Decision and Information Technologies (CoDIT) (2014)
6. Narr, S., Hifulfenhaus, M., Albayrak, S.: Language-independent twitter sentiment analysis. In: The 5th SNA-KDD Workshop (2011)

Big Data Analytics and Visualization: Finance

Shyam Prabhakar and Larry Maves

1 Introduction

Big data Analytics and data science helps finance in combining business research expertise, scientific processes, quantitative analytics and system infrastructure to distill knowledge and insights from internal/external, structured/unstructured data.

All finance institutions have seen an explosion in their velocity, variety and volume of their internal datasets. New federal regulations requirement require leveraging internal and external data linking: [1] Customer service and transactional level data; [2] Social Media activity analysis (Sentimental aka opinion mining); [3] Real time market feeds; [4] Mobile/Online tracking data. Competitive market has accelerated the need for more targeted and accurate decision making thru: [1] Capture and analysis of new source data; [2] Building and direct use of Predictive models; [3] Requirement to run live simulations (stress testing) of potential market events; and the [4] Extracting and storing of increasingly diverse data in raw form for future analysis.

1.1 How Big Data Analytics Could benefit Finance Industry?

Financial institutions is transforming to data-driven organizations as they acquires, process, and leverages data in real time to create organizational efficiencies, delineate new product opportunities, and facilitate navigating the competitive landscape.

S. Prabhakar (✉) • L. Maves
The University of Alabama at Birmingham, Birmingham, AL, USA
e-mail: Shyam.Prabhakar@regions.com

S.C. Suh, T. Anthony (eds.), *Big Data and Visual Analytics*,
https://doi.org/10.1007/978-3-319-63917-8_13

Big Data analytics is providing solutions to long standing business challenges banking and finance industries around the world.

> "Big data is a collection of data that are structured and unstructured from the sources that are inside and outside (Internet) of a company that represents a source for ongoing discovery and analysis."

Financial industries and banks carries lot of data they collect on day-to-day basis. Each day, they add millions of records to their ocean of data. The main question is how to use this information's for their competitive advantage. In a case study IBM did in the year 2012 on Big Data Analytics and its use in financial industry, they found 71% of the banking and financial markets reported that they use the information for their competitive advantage.

Big Data ecosystem provides ability to store large volume of data, which includes structured and unstructured. The echo system also provides tools to format the data and perform analytics and reporting. Some of the top HADOOP vendors are listed below

- Amazon Elastic MapReduce
- Cloudera CDH Hadoop Distribution
- Hortonworks Data Platform
- MapR Hadoop distribution
- IBM Open Platform
- Microsoft Azure's HDInsight -Cloud based Hadoop Distribution
- Pivotal Big Data suite
- Datameer Professional
- Datastax Enterprise Analytics
- Dell—Cloudera Apache Hadoop Solution

Disclaimer
This list of Hadoop vendors are not categorized based on the order of popularity.

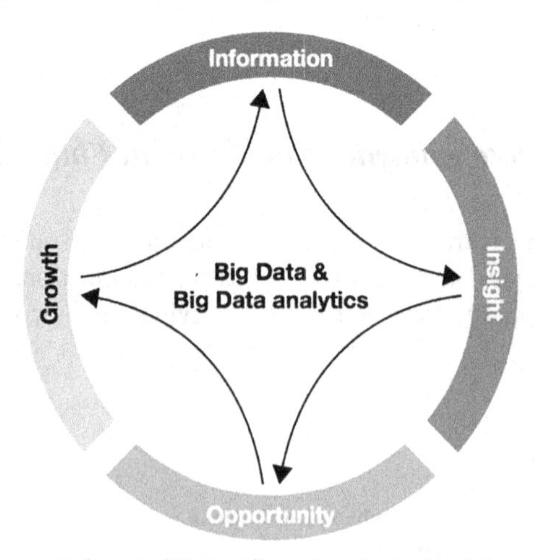

At the root of Big Data lies an important value chain

One of the best and effective big data strategies would be to identify the business requirements first, and then leverage the existing infra-structure, data sources and analytics to support the business opportunity. Almost three-quarters of financial services companies have either started developing a big data strategy or implementing big data as pilot into their process, on par with their cross industry piers.

The promise of achieving business value from big data can only be realized if organizations put into place an information foundation that support the rapidly growing volume, variety and velocity of the data. Inability to connect data across organizational and department silos has been business intelligence challenge for years.

Without powerful analytics software, organizations cannot unlock the true potential of Big Data. Institutions have been using look-back analysis for years and generating trends based on historical data is a common technique. There are new software and tools available for designing and implementing predictive and deep analysis.

2 Areas Where Banking and Financial Institution Could Focus on Using Big Data and Analytics

One of the major area banking and financial industry wants to use Big Data Analytics is for Customer Centric decisions. As majority of the banks and financial institutions market place are moving or forced to move from product centric to customer centric organizations, the customer centric decision-making process is making lot of attention. The data insights, operations, technology and systems are revolving according to customer needs. So understanding a customer behavior and RISK plays important part in the customer analytics/intelligence area. Also by improving the ability to predict changing market and customer preferences, banks and financial institutions could deliver new products and services quickly to seize the opportunities while increasing customer service and loyalty.

Major areas where Big Data Analytics would be of high importance in banking and financial sectors are

- Customer centric decision making
- RISK/financial management—includes prediction of market risk, credit risk etc.
- New Business Model
- Operation optimization

3 Big Data with Machine Learning for Customer Centric Decision Making

Customer centric decision-making allows financial institutions to put customers at the center of all analysis and decision-making, allowing organizations to understand customer behavior and preferences. Knowing your customer helps organization to retain customers and prevent customer churn. Following are some of the steps that should be considered to focus on customer.

1. Organize the data around the customers
2. Put KPI's in place to identify opportunities
   a. How valuable are the new customer an organizations are acquiring?
   b. How are the organizations maximizing the customer relationship?
3. Seek out customer centric insights by focusing on KPI's like CLV(Customer Lifetime Value) and Customer Equity. Also focuses on customer retention specific matrices like the Early repeat rate and Leakey bucket ratio
4. Experiment and Iterate. Experiment the model and iterate to see if better results are achieved with the changing market.

The real value of Big Data for any business is the opportunity to learn about our customers at such depth and speed that we can truly put them in the central stage. Data driven learning enabled by scientific computing and analytic techniques that make it practical to examine data of very high volume, variety and velocity. When these techniques are employed, Big Data yields what is sometimes called Value, the fourth 'V'.

It is no longer enough to have a 360° customer view linking data from all products and channels. Organizations now realize this and adopting more customer centric approach to understand customer needs, understand transaction context, notice what is changing and drive personalized touch points within the window of opportunity. Big Data computing infrastructures are making it practical to employ automated machine learning algorithms for this purpose (Fig. 1).

To start with the customer centric applications, design and automate smart experiments that enable casual predictions. Customer centric prediction models include simple casual predictions of credit risk and line increase to customer retentions. In Dr. Andrew Jennings paper, FICO captures casual relationships in the action-effect models to improve decisions through decision modeling and optimizations (Fig. 2).

With the right algorithms, you can create intelligent applications that can use machine learning to improve data discovery and enhance analytic precision. Value

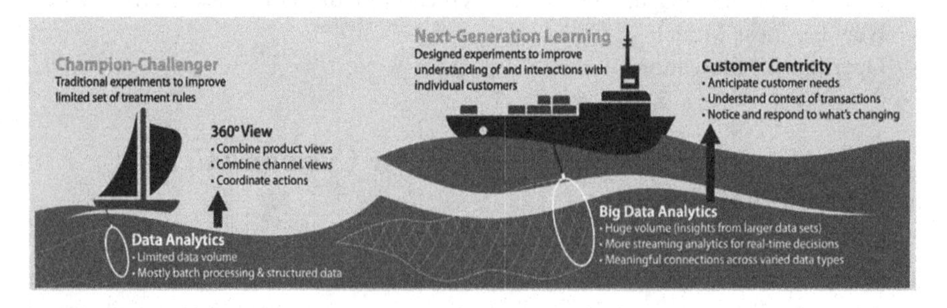

Fig. 1 Courtesy big data way to customer centricity by Dr. Andrew Jennings [5]

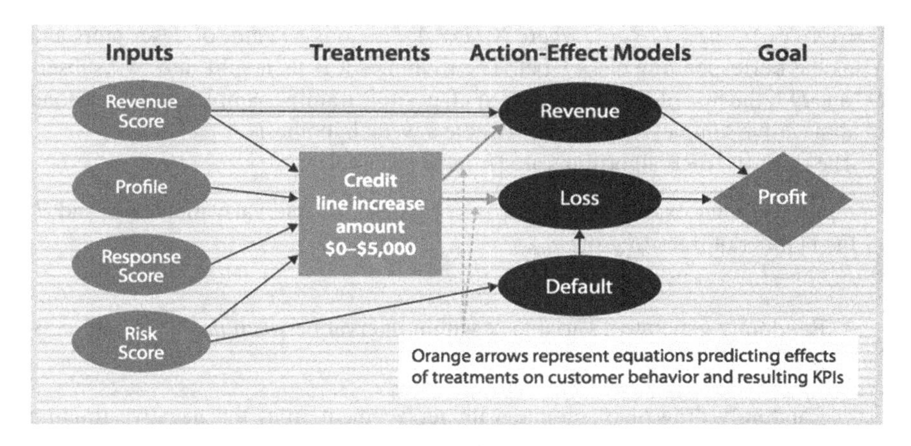

Fig. 2 Courtesy Big data way to Customer centricity by Dr. Andrew Jennings. This shows a model that predicts customer reactions [5]

comes from data-driven insights, not just the data alone. From raw data sources one should analyze all types of data to identify trends, model possible scenarios and predict future results.

Financial institutions like any other, would like to retain their respective customers. In this section, I would like to give an example how big data analytics along with machine learning algorithm help in predicting customer churn. Customer "churn" is defined as the propensity of customers to cease doing business with a company in a given time period. It is very important for a business to know their customers. This helps in providing a high level of customer service and also helps in determining the risk factors. The three main characteristics of churns are (1) the data is usually imbalanced; (2) large learning applications will inevitable have some type of data noise; (3) the task of predicting churn requires the ranking of subscribers according to their likelihood to churn.

One of the journal released in 2009 (Customer churn prediction using improved balanced random forests) recommended Improved Balanced Random Forests (IBRF) is the one predicted more accurately the churn. If you ask any seasoned modelers, they wont recommend one specific method to use, as we have to test with multiple algorithms to find accuracy that could give result in the range of 80–85%. Some of the algorithm tested in this journal were Decision-tree-based algorithm, neural network algorithm, Bayesian multi-net classifier, SVM, sequential patterns and survival analysis have made good attempts to predict churn, but the results were unsatisfactory. The study contributes to the existing literature not only investigating the effectiveness of the random forest approach in predicting customer churn but also by integrating sampling techniques and cost-sensitive learning into random forest to achieve better performance than existing algorithms.

The strategy of random forest is to select randomly subset of m_{try} descriptors to grow trees, each tree being grown in a bootstrap sample of the training set. This number, $\mathrm{m}_{\mathrm{try}}$, is used to split the nodes and is much smaller than the total number if descriptors available for analysis. Algorithm takes as input a training set

$D=\{(X_{1,}Y_{1,}) \ldots .,(X_{n,}Y_{n,})\}$, where $X_{i,}$ $I = 1, \ldots..,n$ is a vector of descriptors and $Y_{i,}$ is the corresponding class label. The training set is then split into two subsets D^{+} and D^{-}, the first of which consists of all positive training samples and the second of all negative samples. Let h_t: $\mapsto\mathbb{R}$ denote a weak hypothesis.

The steps of IBRF algorithms are (from the journal)

- *Input*: Training examples $\{(X_{1,}Y_{1,}) \ldots .,(X_{n,}Y_{n,})\}$; interval variables m and d; the number of trees to grow n_{tree};
- *For* $t = 1, \ldots.., n_{tree}$:
  - Randomly generate a variable $\propto$ within interval between $m - {}^{d}/_{2}$ and $m + {}^{d}/_{2}$;
  - Randomly draw n $\propto$ sample with replacement from the negative training dataset D^{-} and n(1–$\propto$) sample with replacement from the positive training dataset D^{+};
  - Assign w_1 to the negative class and w_2 to the positive class where $w_1 = 1 - \propto$ and $w_2 = \propto$;
  - Grow and unpruned classification tree;
- *Output* the final ranking. Order all test samples by negative scores of each sample. The negative score of each sample can be considered to be the total number of trees, which predict the sample of negative. The more trees predict the sample to be negative, the higher negative score the sample gets.

In this method, the normalized vote for class j at X_1 equals

$$\frac{\sum_{k} I\left(h\left(X_{i,}\right) = j\right) w_k}{\sum_{k} w_k}$$

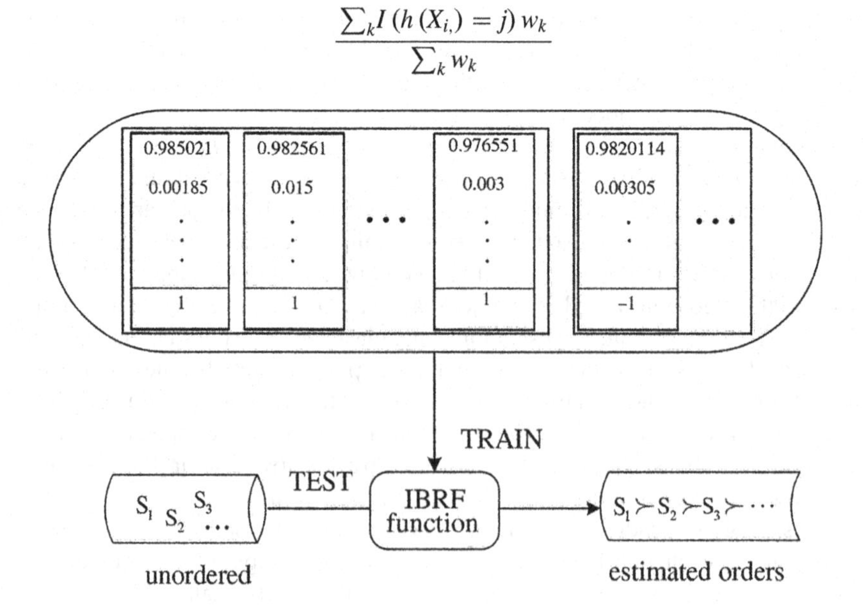

From the sample above, Big Data analytics along with machine learning and NLP could contribute high value in knowing customers, that includes behavior, churn, risk, recommending new products etc.

4 Big Data for Sentiment Analysis Using Natural Language Programming

Big data environment is known for ingesting "variety" of data, which is one of the "V"s along with Volume and Velocity. Big Data platform (Hadoop echo system) is ideal for downloading all the twitter, Facebook and other social media data that could be used to perform sentiment analysis. Main areas in financial industry the sentiment analytics could be used in a financial industry would be able to, (1) identify for stock market variations with prediction ability; (2) Individual financial industries customer experiences over social media; and (3) Brand monitoring and campaign.

Sentiment Analysis is also referred to as Opinion Mining on Big Data is a current topic for research as this leverage an opportunity to use already present data in open and analyze the same. Researchers have been interested in automatically detecting sentiment in texts for at least a decade. We can label some texts as positive, some texts as negative and some as normal and use these to train and algorithm that detects sentiment in texts. This information could lead investors and financial decision makers to make strategic decisions on the impact the market. The financial data from news, blogs and social medias are noisier than other segments due to changing market trends.

As per one of the published journal by Harvard College Cambridge, Massachusetts in the year 2009 recommends two different ways we can analyze sentiments. (1) Using NLP (Natural Language Programming), in which the words that are relevant to sentiment are automatically learned from the data. Each new story is summarized as a binary vector $(w_1, \ldots, w_n)$. The attribute w_1 is equal to 1 if word i present in the story and equal to zero if word i is absent; (2) Non-linear algorithms such as decision trees and support vector machines are frequently used to map the words in a text to its classification. There has been on going research work in this area to find the sentiment analysis in financial markets.

Sentiment Analysis can be carried out in two ways namely supervised and unsupervised. The supervised learning requires a training dataset and assumes a finite number of categories for an item to fall into. Some of the supervised learning algorithms we could use here are Naïve Bayes, SVM or KNN etc. Unsupervised learning is based on getting Semantic Orientation (SO) of the document. Sentiment Analysis also needs to have a dictionary for comparing bag of words.

Sentiment Analysis depends upon the quality of the text used and training data sets. There are other methods adopted by different studies, for example, dictionary based technique. This method uses dictionary of sentiment bearing words to classify

Table 1 Polarity levels

Category	Polarity	Display
1	Positive	Excited
2	Strong positive	Happy
3	Negative	Upset
4	Strong negative	Angry

texts into positive, negative or neutral options. The results from this may not have the same accuracy when compared to machine learning based algorithms. Some of the main areas in financial that we could use NLP are listed below (courtesy https://www.ibm.com/blogs/watson/2016/06/natural-language-processing-transforming-financial-industry-2/)

- Gather real-time intelligence on specific stocks
- Provide key hire alerts
- Monitor company sentiment
- Anticipate client concerns
- Upgrade quality of analyst reporting
- Understand respond to news events
- Detect insider trading

5 Big Data Analytics and Business Intelligence

Business Intelligence is a process from data acquisition to data presentation where the audience looks at the presented data for easy, well-defined reports and answers. Big Analytics helps you find questions you don't know you want to ask.

Big data analytics is the process of examining large (Volume), varied data sets (Variety) which is loaded fast into the disk (Velocity) to uncover patterns, unknown correlations, market trends, customer preferences and other useful information that can help organizations to make more informed business decisions. Big Data analytics enable data scientist, predictive modelers, statisticians and other analytics professionals to analyze high volumes of structured transactional data and other forms of data often left untapped by conventional business intelligence process. This includes unstructured and semi-structured data as well. For example, internet click stream data, web server logs, social media contents, text from customer emails, survey responses, mobile-phone call-detail records and machine data captured by sensors connect to the internet of things (IoT). Big data analytics is a form of advanced analytics, which involves complex applications with elements such as predictive models, statistical algorithms and what-if analyses powered by high-performance analytics systems.

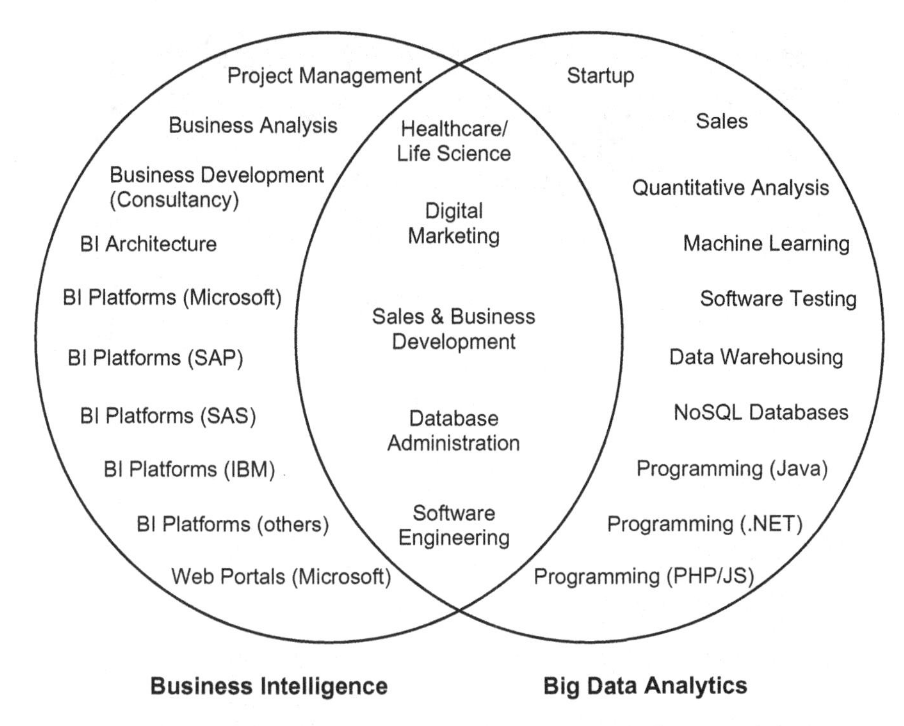

By combining the power of big data analytics and use of BI may bring another area that could use the best of both to determine business intuitions. Lot of studies suggests that business intelligence need to adapt to big data (Fig. 3).

6 Conclusion

Big data used to be a technical problem but now it is a business opportunity for lot of businesses. A 2015 research indicated that 63% of the firms using BI indicate that Big Data analytics is created by/used by departments other than IT. Big data has diverse data types, delivered at various speed and frequencies. These diverse data combine with an organizations transaction and other structured data and use collection different tools for predictive analytics, data mining, statistics, artificial intelligence, natural language processing, machine learning etc. provides you with Big Data Analytics. Big data analytics explores granular details of business operations and customer interactions that seldom find their way into a data warehouse of standard BI reports.

Big data analytics and visualization brings a new culture of decision-making. The evidence is clear that the data-driven decisions give you better decisions. Leaders will either embrace this fact or will be replaced. It is also clear that the companies that figure out how to combine domain expertise with data science will pull away from their rivals.

Fig. 3 A pictorial representation of why big data analytics is important and how it differs from BI [5]

References

1. IBM institute for business value: Analytics: The real-world use of big data in financial services: https://www.935.ibm.com/services/multimedia/Analytics_The_real_world_use_of_big_data_in_Financial_services_Mai_2013.pdf
2. Big data analytics: What it is and why it matters? SAS https://www.sas.com/en_us/insights/analytics/big-data-analytics.html
3. IBM How NLP is transforming financial industry: https://www.ibm.com/blogs/watson/2016/06/natural-language-processing-transforming-financial-industry-2/
4. Customer churn prediction using improved balanced random forests by Yaya Xie A, Xiu Li A, E.W.T. Ngai, Weiyun Ying C https://pdfs.semanticscholar.org/4359/d76d3b36944553eb1d08befaf219122fbefd.pdf
5. FICO: When is Big Data the way to customer centricity? Next generation analytic learning finds critical insights in an ocean of false clues by Dr. Andrews Jennings; http://www.fico.com/landing/pdf/67_Big_Data_Customer_Centricity_2951WP.pdf

Study of Hardware Trojans in a Closed Loop Control System for an Internet-of-Things Application

Ranveer Kumar and Karthikeyan Lingasubramanian

Abstract A closed-loop system is a primary technology used to automate our critical infrastructure and major industries to improve their efficiency. Their dependability is challenged by probable vulnerabilities in the core computing system. These vulnerabilities can appear on both front (software) and back (hardware) ends of the computing system. While the software vulnerabilities are well researched and documented, the hardware ones are normally overlooked. However, with hardware-inclusive technological evolutions like Cyber-Physical Systems and Internet-of-Things, hardware vulnerabilities should be addressed appropriately. In this work, we present a study of one such vulnerability, called Hardware Trojan (HT), on a closed-loop control system. Since a typical hardware Trojan is a small and stealthy digital circuit, we present a test platform built using FPGA-in-the-loop, where the computing system is represented as a digital hardware. Through this platform, a comprehensive runtime analysis of hardware Trojan is accomplished and we developed four threat models that can lead to destabilization of the closed-loop system causing hazardous conditions. Since the primary objective is to study the effects of hardware Trojans, they are designed in such a way that they can be accessible and controllable.

1 Introduction

A closed-loop system is a primary technology used to automate our critical infrastructure and major industries to improve their efficiency. Their dependability is challenged by probable vulnerabilities in the core computing system. These vulnerabilities can appear on both front (software) and back (hardware) ends of the computing system. While the software vulnerabilities are well researched and documented, the hardware ones are normally overlooked. However, with hardware-inclusive technological evolutions like Cyber-Physical Systems and Internet-of-Things, hardware vulnerabilities should be addressed appropriately. In this work,

R. Kumar (✉) • K. Lingasubramanian
The University of Alabama at Birmingham, Birmingham, AL35294, USA
e-mail: ranveerk@uab.edu

S.C. Suh, T. Anthony (eds.), *Big Data and Visual Analytics*,
https://doi.org/10.1007/978-3-319-63917-8_14

we present a study of one such vulnerability, called Hardware Trojan (HT), on a closed-loop control system. Since a typical hardware Trojan is a small and stealthy digital circuit, we present a test platform built using FPGA-in-the-loop, where the computing system is represented as a digital hardware. Through this platform, a comprehensive runtime analysis of hardware Trojan is accomplished and we developed four threat models that can lead to destabilization of the closed-loop system causing hazardous conditions. Since the primary objective is to study the effects of hardware Trojans, they are designed in such a way that they can be accessible and controllable.

A PID controller is the most widely used controller in the industry due to its multiple advantages. FPGAs are also nowadays widely used in industry to fulfill high frequency requirements. Multiple PIDS are implemented on the FPGA [1]. Controllers can control multiple sensors. Due to a rapid increase in the use of FPGA based PID controllers, the possibility of threats are also increased. People are more focused and aware towards software based viruses or Trojans but less aware of hardware Trojans. So, there is need to explore the possibility of hardware Trojans in programmable IoT devices and how it can affect closed loop control systems.

Hardware Trojans are the malicious circuits added in the Integrated Circuits (ICs) which can affect either the functionality or leak information of the ICs or the devices. In order to reduce the manufacturing cost of ICs, companies fabricate their ICs in outside foundries, which makes the ICs vulnerable to the attacks during various stages of their lifecycle [2]. Hardware Trojan added in the original circuits can introduce fault in the chip or provide access to the attackers. It can also leak secret information or disable the targeted circuit or device [3]. Karri, Rajendran, Rosenfeld, and Tehranipoor proposed that the chips are vulnerable to HT at various levels of IC development cycle including specification, design, fabrication, testing, and assembly. HT are also inserted at various hardware abstraction levels including system level, RTL level, gate level, transistor level, and physical level [4]. IoT devices that are controlled by these chips are vulnerable to intrusions and attacks making hardware security an important area of research since an attacker could potentially gain control of benign devices such a sensors that monitor a homes temperature to more controllable devices such as the homes thermostat or door locks.

A Kill Switch is an extra circuitry present in the chip, which can disrupt the functioning of the chip and cause the chip to die. It can be inserted in the chip during the manufacturing or the design phase. There were multiple incidents, which made researchers and the world to think about Kill Switch. In September 2007, Israeli jets bombed a nuclear installation in northeastern Syria. Some might doubt that the incident was a result of a fabricated hidden "backdoor" in the Syrian Radar, which either disrupted the chip's function or temporarily blocked the radar [5]. Most of the chips are manufactured in foreign foundries and not all are secure which can cause software as well as hardware attacks. Fighter jets can contain thousands of chips, which makes it enormously vulnerable to any extraneous circuitry. According to Professor Ruby Lee, Cryptographic Expert, Electrical Engineering, Princeton, "you won't check for the infinite possible things which are not specified" [5]. Tests are not reliable enough to identify any extraneous circuitry which are not active and do not affect the normal functionality of the circuit. Verification and testing cost also were increased depending on the number of chips being tested. Verification engineers randomly select a chip from thousands for physical inspections. More than 90 percent of FPGAs are fabricated in foundries outside the United States and there is no way to guarantee that they have not been tampered with [5].

2 Hardware Trojan Detection and Emerging Solutions

According to Yier Jin, standard testing methods are not sufficient to detect hardware Trojans due to following reasons [6]:

1. The behavior of the Trojan is unexpected and structural pattern testing may not cover Trojan test vectors.
2. It is hard to predict the functionality of the Trojan. Routine functional testing cannot disclose the malicious hidden extra circuitry and function.
3. Considering all input patterns while testing is not possible for complex chips.

   Destructive reverse engineering is effective with high cost to check the integrity of the chips [6].

There are three common basic approaches used for detection of Trojans [7]:

1. Failure Analysis-Based Techniques

   This technique includes scanning electron microscopy (SEM), pico-second imaging circuit analysis (PICA), voltage contrast imaging (VCI), light-induced voltage alteration (LIVA), charge induced voltage alteration (CIVA), etc. [8]. This technique is very cost effective and time consuming and it is not possible to check all the chips while the Trojans might be present in random chips.
2. ATPG-based Trojan Detection Techniques

   This technique includes detection of a Trojan by applying digital stimuli, which can be obtained from the netlist of the chip and analyze the output of the chip.
3. Side Channel Signal Analysis

   This technique includes power analysis, electromagnetic field analysis, temperature variations, voltage variations, timing analysis, etc. According to Wang and Jim, Side-Channel-Based analysis can be improved with design-for-hardware-trust (DFHT).

3 Design and Implementation

3.1 Discrete-Time Control System

A controller is a kind of a digital computer [9]. Time sharing can enable a controller to control multiple functions. For instance, the space shuttle main engine (SSME)

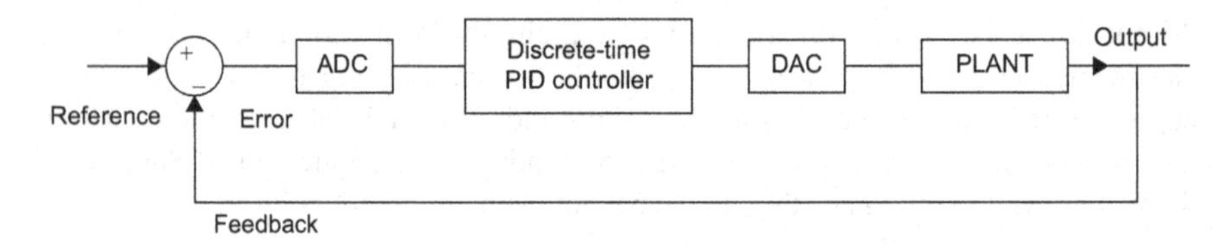

Fig. 1 Closed loop control system with discrete-time PID controller

controller has two digital controllers, which control pressure, temperature, flow rates, turbo pump speed, valve positions, and engine servo-valve actuator positions [10] (Fig. 1).

This closed loop system consists of four main components.

1. Plant: continuous-time dynamic system
2. ADC converter
3. DAC converter
4. Discrete PID Controller

3.2 Discrete PID Controller

A PID is a three-term controller where *P* stands for proportional, *I* stands for integral, and *D* stands for the derivative components in the controller. A PID is most widely used in industrial control systems. k_P is proportional gain, k_I is integral gain, and k_D is deferential gain. Standard discrete-time PID controller [11] can be represented as

$$C = K_P\left(1 + \frac{1}{T_i}IF(z) + \frac{T_d}{\frac{T_d}{N} + DF(z)}\right)$$

$$IF(z) = \frac{T_S}{2}\frac{z+1}{z-1}$$

$$DF(z) = \frac{T_S z}{z-1}$$

where, K_P is proportional gain, T_i is integrator time, T_d is derivative time, N = derivative filter divisor. IF(z) is a discrete integrator formula for the integrator filter, and DF(z) is a discrete integrator formula for derivative filter.

Primary objectives of control system analysis and design [10].

1. Producing the desired transient response: It affects the speed of the system and influences human patience and comfort.

2. Reducing steady-state error: Steady state responds to the accuracy of the control system. It governs how closely output matches the desired response.
3. Achieving stability: A system must be stable to produce the proper transient and steady state response.

3.3 Hardware Trojan Taxonomy

According to Karri, Rajendran, Rosenfeld, and Tehranipoor, the taxonomy of hardware Trojan has six different categories after implementing numerous hardware Trojans [4]. The figure explains the different categories of taxonomy of hardware Trojans implemented in our case study. Hardware Trojans are inserted in the design phase in the form of VHDL code at RTL level. We used time-based internally triggered hardware Trojans, which can change the functionality of the system. Hardware Trojans stay hidden in the processor or the computational core of the system. Insertion of hardware Trojans changes the layout of the design and can change the size of the design or the programmable bit-stream (Fig. 2).

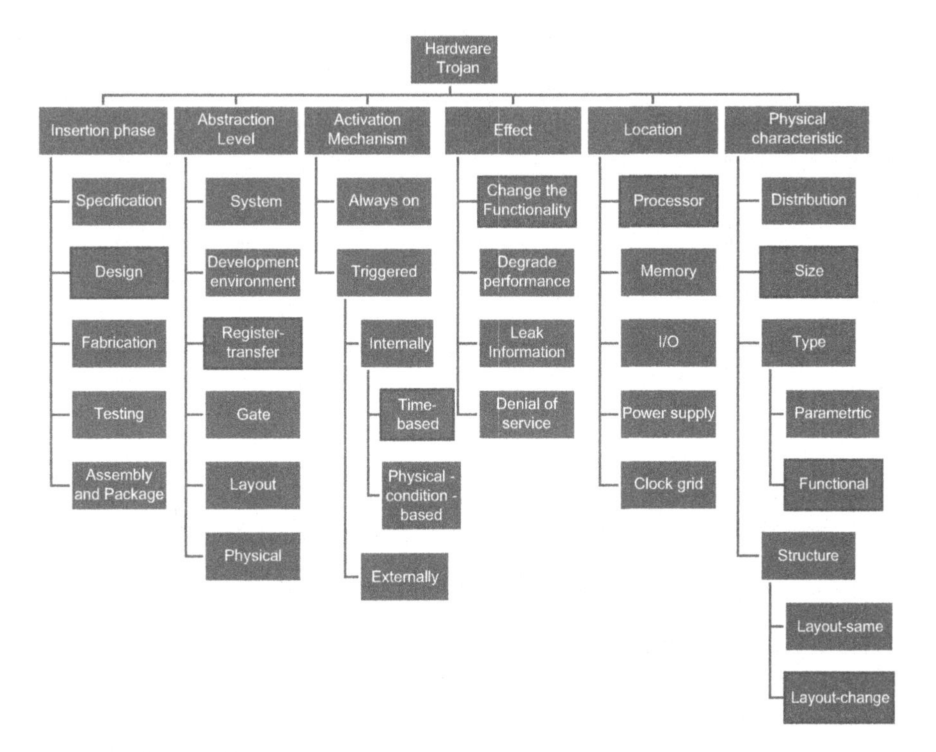

Fig. 2 Hardware trojan taxonomy

3.4 *Threat Model Implementation in a Closed Loop Control System*

The hardware Trojans are inserted in the form of VHDL code in the discrete PID controller and simulations were performed in an FPGA environment. A HDL Coder is used to generate the VHDL code for the discrete PID controller, and hardware Trojans in the form of VHDL are inserted. A HDL Verifier is used to compile the VHDL codes and generate programmable bit-stream to program FPGA and perform FPGA-in-the-Loop Simulation (Figs. 3 and 4).

The following is the digital implementation of PID controller using Quartus Prime RTL viewer to generate the image shown below.

The hardware Trojan is inserted in the PID controller. We implemented a time-based internally triggered hardware Trojan. Here, we used two 32-bit counters to specify the trigger time of the hardware Trojan. The counter increases incrementally on the rising edge of the clock. Once the counter value matches the specified count value, it sets the trigger value equal to 1. It is highlighted in the RTL view of the PID controller (Fig. 5).

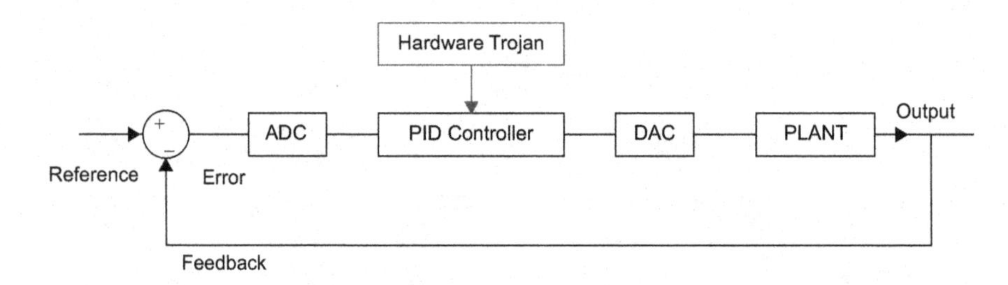

Fig. 3 Hardware trojan in PID controller

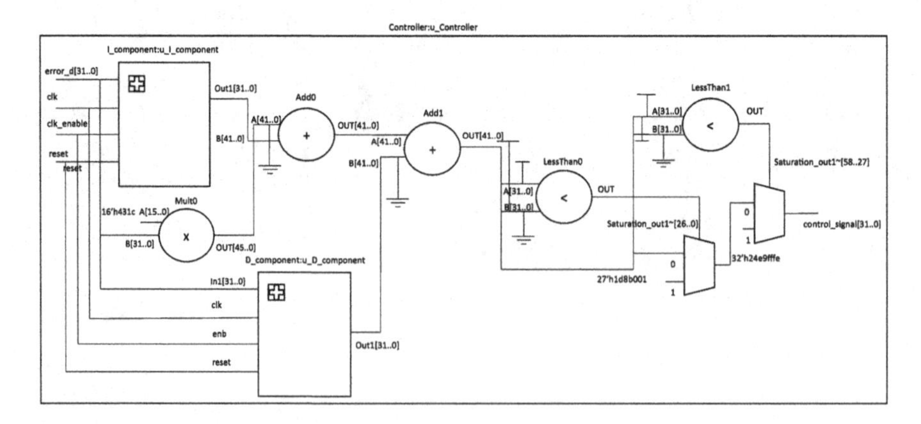

Fig. 4 Digital implementation of PID controller

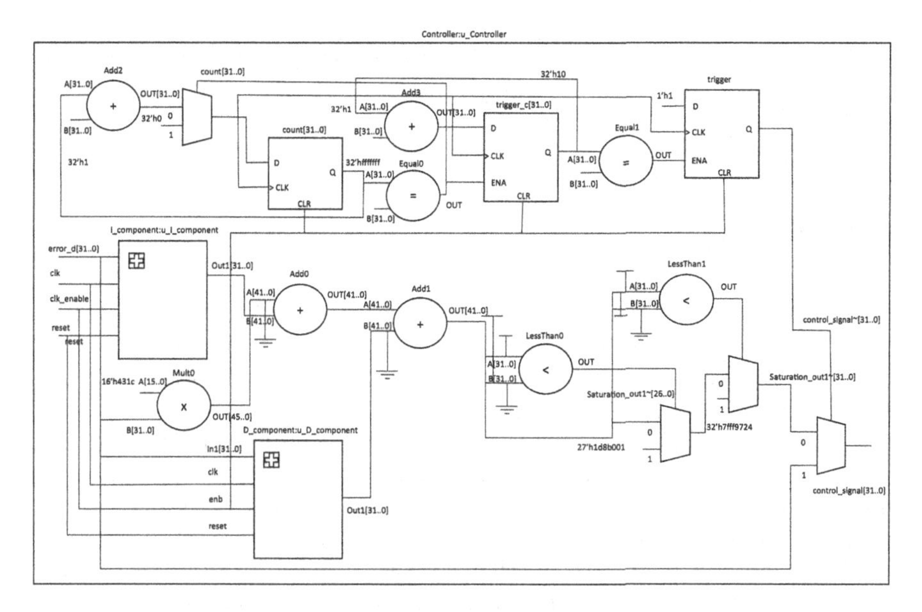

Fig. 5 Digital Implementation of PID Controller with Hardware Trojan

3.5 Experimental Setup

The Equipment used for the FPGA-in-the-Loop simulation are Computer - Intel Core 2 duo processor is used in this work. Multiprocessor CPU can be used to make the compilation of the VHDL code faster. We can also define the number of processors used in the compilation process of the VHDL code using Intel Quartus Prime.

FPGA - BeMicro CVA9 Cyclone V FPGA kit is used in this work. A JTAG is used for the connection of FPGA with the computer to load the bit file and perform FPGA-in-the-Loop. Some of the software tools used are MATLAB R2016b, Simulink, HDL Coder, HDL Verifier and Intel Quartus Prime 15.1.

HDL Coder is used to generate portable, synthesizable Verilog/VHDL code from MATLAB functions, State-flow charts and Simulink models [12]. HDL Coder has specific library which include over 200 blocks and state flow charts, which are supported by HDL Coder. hdllib command can be used to filter the Simulink library and to get HDL Coder supported blocks. These blocks should be used for modeling the logic, subsystem or system. The generated code is target independent and can be used in Quartus Prime or Xilinx tool chain. The generated HDL code can be used for FPGA programming or ASIC prototyping and design without writing a single line of code. We can also edit the generated HDL Codes for different purpose. Altera and Xilinx FPGAs can be programmed using the inbuilt workflow advisor in Simulink. We can add custom FPGAs and program it. We can save our time by using HDL Coder to generate Verilog/VHDL codes of complex systems.

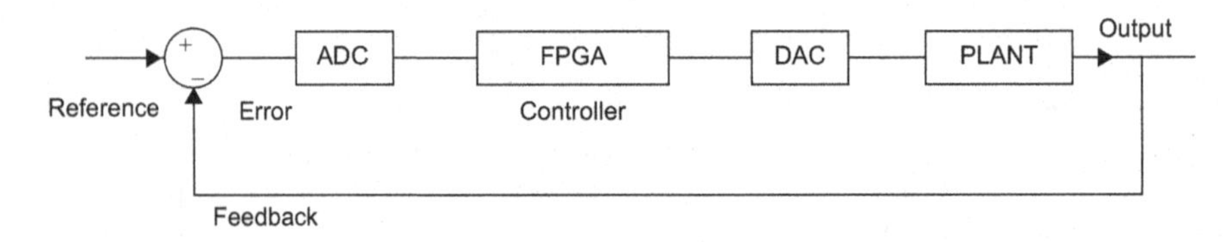

Fig. 6 FPGA-in-the-loop

HDL Verifier is an add-on tool of Simulink provided by MathWorks. It is used to automatically generate test benches and to verify Verilog/VHDL design using available verification wizards. HDL Verifier is used for Co-simulation and FPGA-in-the-loop using Xilinx and Altera FPGAs. Co-simulation allows users to simulate the design with MATALB/MATLAB System Object/SIMULINK as well as on ModelSim or Incisive Simulator from Cadence. FPGA-in-the-Loop is used to generate equivalent block to replace with the Simulink block and generate programmable bitsream to program the FPGA. HDL Verifier uses third party tool for the compilation of the Verilog/VHDL code and generate bit-stream to program Xilinx or Altera FPGAs. It also eliminates the need of writing independent test benches for the verification of the design (Fig. 6).

Intel Quartus prime version 15.1 is used by HDL Verifier for the compilation of the VHDL code. It will run automatically in the background of HDL Verifier FPGA-in-the-Loop process.

3.6 Design Workflow

The above is the design flow, which is followed while implementation. We create our model using Simulink and or MATLAB. Then, we select the block, which needs to be implemented on FPGA. We need to check the compatibility of the selected block with HDL Coder. Portable and synthesizable Verilog/VHDL code is generated using HDL Coder. Hardware Trojans might be designed independently and inserted into the generated Verilog/VHDL code (Fig. 7).

HDL Verifier is used to perform FPGA-in-the-Loop Simulation. An equivalent block is generated suing the VHDL code and the previous block is then replaced by the generated block having HT. Intel Quartus is used in the back end to compile and generate the programmable *sof* file. We can set the simulation time and load the generated bit file to the connected FPGA. We used scope and data inspector to visualize the simulation result and the hardware Trojan effect.

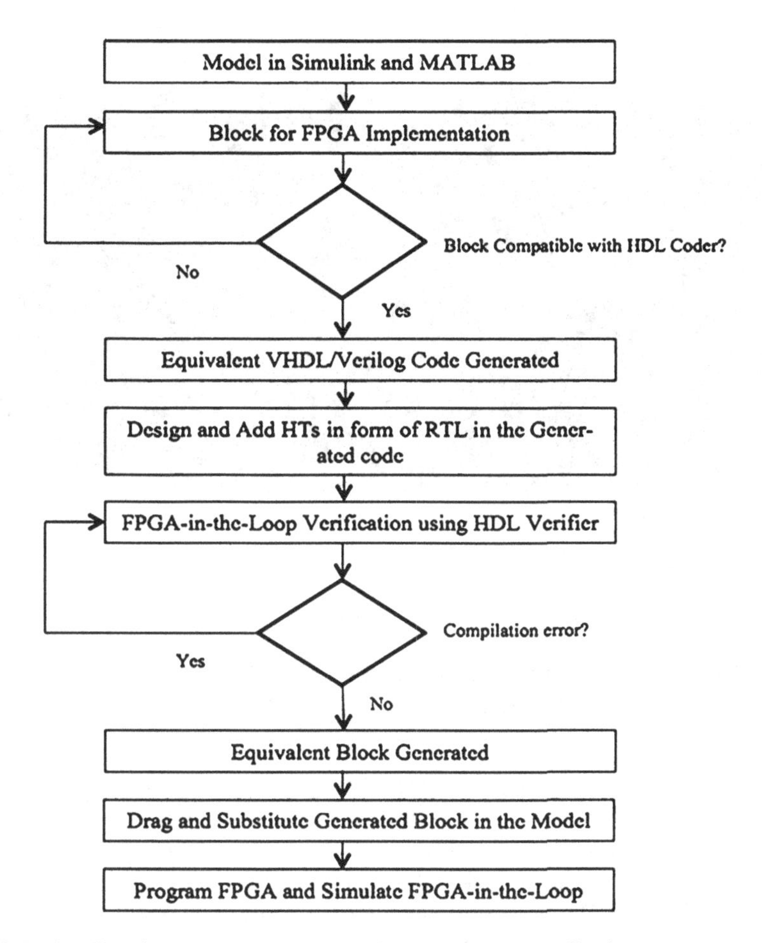

Fig. 7 Design Flow for studying Hardware Trojans in IoT device applications

4 Results

4.1 Threat Model Implementation in PID Controller: Threat I—Turn Off Controller

In this threat model, we are introducing trigger based hardware Trojan which removes the controller from the closed loop system. Once the hardware Trojan get triggered at the specified count value, it removes the controller from the closed loop and system is not able to reach the steady and state and remove error. It becomes unstable and finally break after sometime. We can always select the trigger time of the hardware Trojan while designing and introducing it in the VHDL code of the PID controller (Fig. 8).

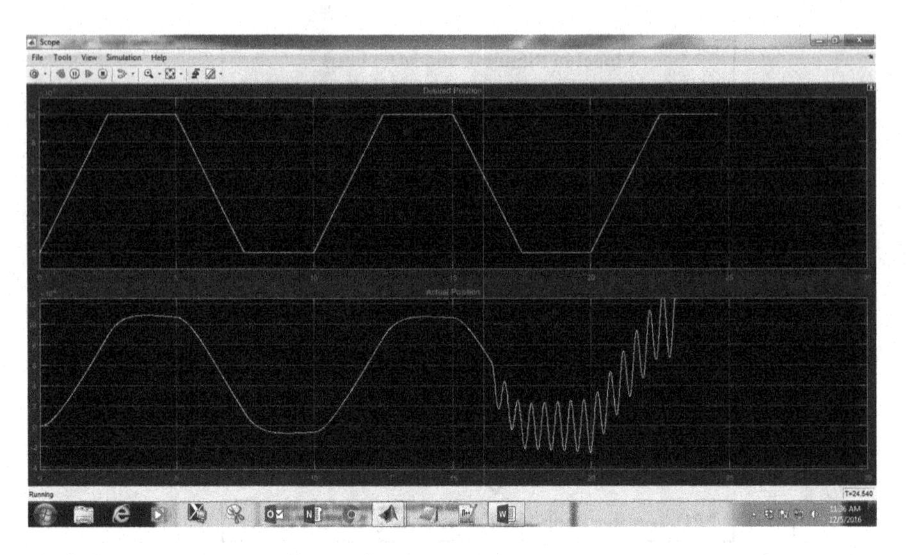

Fig. 8 Threat model 1: turn off controller

4.2 Threat Model Implementation in PID Controller: Threat II—Turn Off Controller for a Short Time

In this threat model, we introduced trigger based HT which removes the controller from the closed loop at time t = 18 s and put the controller back after a very short time t = 21 s into the closed loop system. We observed that removing the controller from the closed loop system makes the signal unstable, but the system gains the stability. In this case, HT don't affect the system. But it will be destructive for timing critical closed loop control systems (Fig. 9).

4.3 Threat Model Implementation in PID Controller: Threat III: Turn Off and On Controller after 17 S

In this threat model, we introduced trigger based HT which removes the controller from the closed loop at time t = 18 s and again putting the controller back at t = 35 s into the closed loop system. We can see that taking out the controller from the closed loop system makes the system unstable and there is a sharp rise in the amplitude of the pulse in y-axis. When the controller again activated in the system it makes the system stable and the system behaves as expected but at different amplitude (Fig. 10).

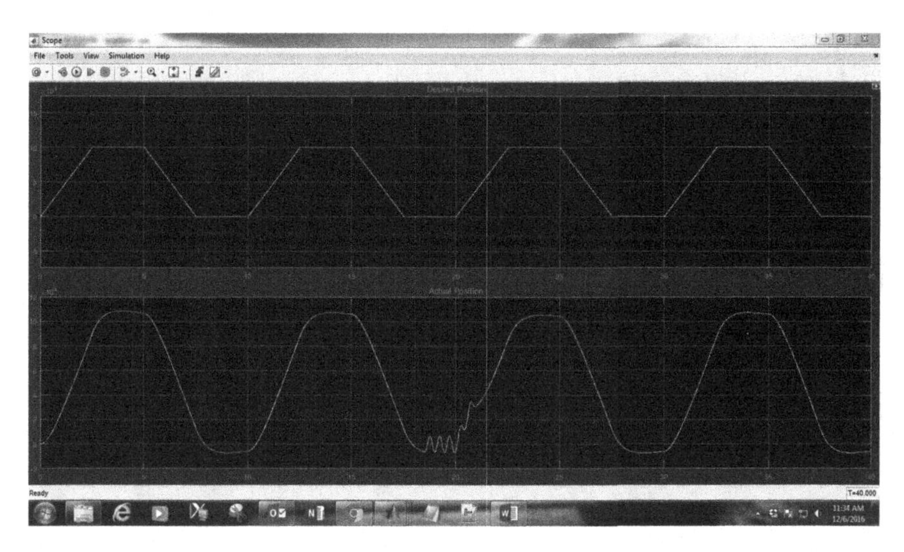

Fig. 9 Threat model 2: turn off controller for a short time

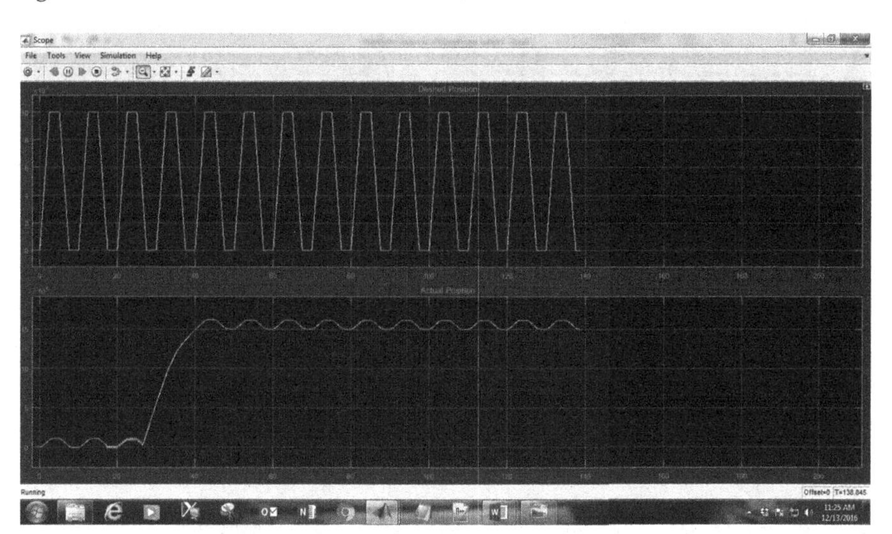

Fig. 10 Threat Model 3: Turn off and on controller after 17 s

4.4 Threat Model Implementation in PID Controller: Threat IV—Delay in Controller: Variable Delay Length, and Threshold Delay vs No-Delay

In this threat model, Internally trigger based HT is used, which introduces delay (actuator delay) in the control system and can breaks the system after sometime. The delay length is variable. We simulated the FPGA based PID controller for various

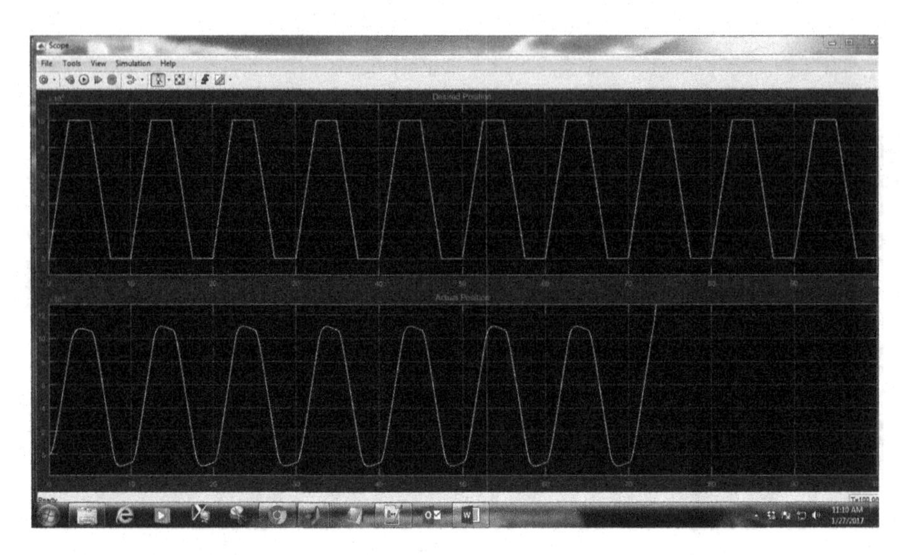

Fig. 11 Threat model 4: delay in controller

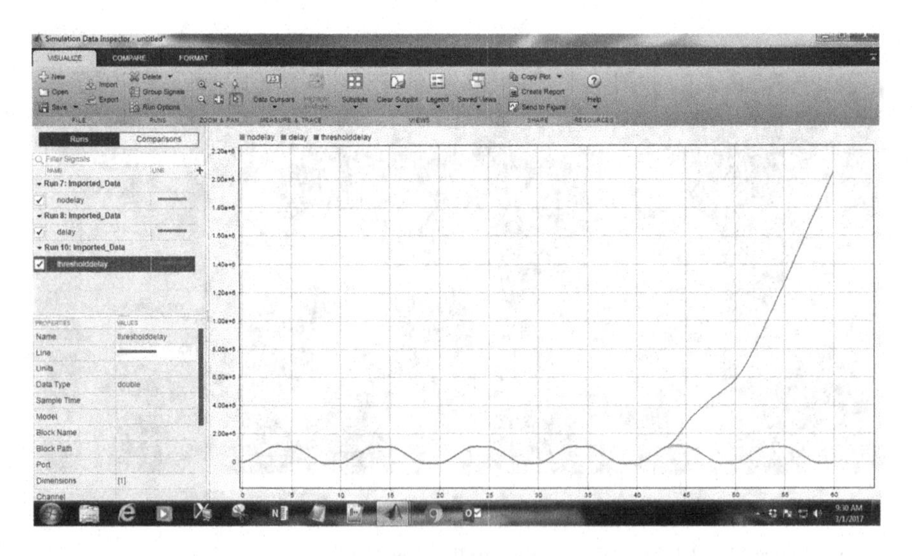

Fig. 12 Threshold delay v. delay v. no delay

values of delay. For delay value 196, the closed loop control systems get unstable and break the system (Figs. 11 and 12).

For delay lengths less than 196, we can the unstability and the variations in the output of the system. The above result include three signal, which shows the comparison between the output of the Closed loop control system having threshold delay (196), delay and no delay. Here, Threshold delay is that value of delay for which this closed loop control system become unstable and break the system.

5 Conclusion

Closed-loop systems are the stepping-stone towards emerging technological innovations based on automation. It is important to understand its various vulnerabilities in order to achieve comprehensive security of our cyber infrastructure. In this work, we have presented a FPGA based experimentation platform to study the effect of hardware Trojans on Closed-loop systems. We implanted hardware Trojan on the digital controller of closed-loop system and developed four threat models, (1) permanently disabling the controller, (2) disabling the controller for shorter period of time, (3) disabling the controller for longer period of time, (4) introducing additional delay elements in the controller. This work can be improved by studying more threat models based on versatile inclusion of delay elements. In addition, the study can be improved with more realistic modeling of components, with the hardware Trojan stealthily embedded with internal trigger channel. Such a model can act as a test platform to investigate potential countermeasures.

References

1. Aboelaze, M., Shehata, M.G.: Implementation of multiple PID controllers on FPGA. In: 2015 IEEE International Conference on Electronics, Circuits, and Systems (ICECS), Cairo, pp. 446–449 (2015). doi: https://doi.org/10.1109/ICECS.2015.7440344
2. Shila, D.M., Venugopal, V.: Design, implementation and security analysis of hardware trojan threats in FPGA. In: Communication and Information Systems Security Symposium, IEEE ICC (2014)
3. Lv, Y.Q., Zhou, Q., Cai, Y.C.: Trusted integrated circuits: the problem and challenges. J. Comput. Sci. Technol. **29**(5), 918–928 (2014). https://doi.org/10.1007/s11390-014-1479-9
4. Karri, R., Rajendran, J., Rosenfeld, K., Tehranipoor, M.: Trustworthy hardware: identifying and classifying hardware trojans. Computer. **43**(10), 39–46 (2010)
5. Adee, S.: the hunt for the kill switch. Spectrum IEEE. **45**(5), 34–39 (2008)
6. Jin, Y., Kupp, N., Makris, Y.: Experiences in hardware trojan design and implementation. In: 2009 IEEE International Workshop on Hardware-Oriented Security and Trust, Francisco, CA pp. 50–57
7. Wang, X., Tehranipoor, M., Plusquellic, J.: Detecting malicious inclusions in secure hardware: challenges and solutions. In: Proceedings of the 2008 IEEE International Workshop on Hardware-Oriented Security and Trust, HST'08, IEEE Computer Society, Washington, DC, USA, pp. 15–19 (2008)
8. Soden, J., Anderson, R., Henderson, C.: IC failure analysis tools and techniques – magic, mystery, and science. In: International Test Conference, Lecture Series II "Practical Aspects of IC Diagnosis and Failure Analysis: A Walk through the Process", pp. 1–11. (1996)
9. Online, http://www.idsc.ethz.ch/content/dam/ethz/special-interest/mavt/dynamic-systems-n-control/idsc-dam/Lectures/Control-Systems-2/RT2_skript_digitale_regelung_FS2015.pdf
10. Nise Norman, S.: Control system engineering, 3rd edn. John Wiley & Sons, Inc, New York, NY (2000)
11. MathWorks, https://www.mathworks.com/help/control/ug/discrete-time-proportional-integral-derivative-pid-controller.html
12. MathWorks, HDL Coder: https://www.mathworks.com/products/hdl-coder.html

High Performance/Throughput Computing Workflow for a Neuro-Imaging Application: Provenance and Approaches

T. Anthony, J.P. Robinson, J.R. Marstrander, G.R. Brook, M. Horton, and F.M. Skidmore

Abstract We describe a high performance/throughput computing approach for a full-brain bootstrapped analysis of Diffusion Tensor Imaging (DTI), with a targeted goal of robustly differentiating individuals with Parkinson's Disease (PD) from healthy adults without PD. Individual brains vary substantially in size and shape, and may even vary structurally (particularly in the case of brain disease). This variability poses significant challenges in extracting diagnostically relevant information from Magnetic Resonance (MR) imaging as brain structures in raw images are typically very poorly aligned. Moreover, these misalignments are poorly captured by simple alignment procedures (such as whole image 12-parameter affine procedures). Non-linear warping procedures that are computationally expensive are often required. Subsequent to warping, intensive statistical bootstrapping procedures (also computationally expensive) may further be required for some purposes, such as generating classifiers. We show that distributing the preprocessing of the images using a compute cluster and running multiple preprocessings in parallel can substantially reduced the time required for the images to be ready for quality control and the final bootstrapped analysis. The proposed pipeline was very effective developing classifiers for individual prediction that are resilient in the face of inter-subject variability, while reducing the time required for the analysis from a few months or years to a few weeks.

1 Introduction

Medical research and practice has been faced with an exponential growth in information that is stored in digital formats. This large volume of stored information presents both opportunities and challenges. Magnetic Resonance Imaging (MRI)

T. Anthony (✉) • J.P. Robinson • J.R. Marstrander • F.M. Skidmore
University of Alabama at Birmingham, Birmingham, AL, USA
e-mail: tanthony@uab.edu

G.R. Brook • M. Horton
National Institute of Computational Sciences, University of Tennessee, Knoxville, TN, USA

S.C. Suh, T. Anthony (eds.), *Big Data and Visual Analytics*,
https://doi.org/10.1007/978-3-319-63917-8_15

is a particularly prominent (and currently untapped) source of potential medically relevant data. In the United States, there are over 10,000 MRI machines [1], and in 2011 alone, over 8.7 million MRI brain scans occurred. The human brain comprises about 1350 cubic centimeter of tissue, blood vessels, and fluid spaces, surrounded by other tissues such as the skull/scalp, pharynx and larynx, and other structures that must be differentiated from the brain. At common imaging resolution (1 mm $\times$ 1 mm $\times$ 1 mm voxels), a typical 5 min brain scan may easily contain 7.2 million voxels. Since many brain scans contain multiple dimensions of data (such as time), and a 30–40 min scan session typically includes multiple brain scans encoding different brain information, MR has not only a large amount of data, but also embedded high dimensionality on an individual level. Comparing individuals in large group analyses adds even more complexity, as individual brains vary in size, shape, and configuration. Brain imaging is commonly stored and used in clinical practice, where a human reader typically reviews the imaging and provides a written summary of findings. However, the high dimensionality of imaging data means that embedded information pertaining to discrete diseases is not always easily accessible or visible to a human reader, and focused analysis of MR has become a growing research field. Researchers now have access to large, freely available research databases that also include MRI data [2]. It is clear that to make practical use of this massive amount of imaging data, high performance computing solutions will be required. In many disciplines, computer simulations are now used as a means of discovering and validating scientific models. This so-called "in silico" experimentation has been fueled by a number of advances in computational technology, including the ability to archive and distribute massive amounts of data and the availability of shared hardware resources via grid computing.

This paper presents a workflow for processing of MRI images of the human brain, with a particular focus on Parkinson's Disease (PD). Previous research into PD has shown significant group-wise differences between MRI brain images of PD and healthy adults. In particular, differences exist in measures including resting brain activity and blood flow [3, 4], diffusion tensor fractional anisotropy [5], and brain iron deposition [6, 7]. Despite these promising findings, none of these putative biomarkers have become utilized in the field. With further research and increases in computational power, it is hoped that these measures may find wider use in research and clinical practice.

There are many open source neuroimaging analysis toolkits [8–11]. One of the most popular packages is the Analysis of Functional Neuroimaging (AFNI) currently distributed by the NIH. AFNI is a comprehensive toolkit that covers almost every aspect of fMRI image analysis [12–14]. Another toolkit that is used in the workflow described in this paper is TORTOISE [15]. The TORTOISE software package is for processing diffusion MRI data, and consist of three main modules, diff_prep, diff_calc and, dr_buddi. The workflow presented by this paper utilizes the AFNI and TORTOISE software suites in a high-performance computing environment. To make maximal use of available resources and newer computational fabrics, significant optimizations were required to the 3dQwarp from the AFNI suite [16].

2 Methods

The following section shows the workflow for pre-processing images obtained from MR imaging in order to run a robust bootstrapped analysis to differentiate Parkinson's disease from Healthy controls. The following section is further divided into three sub-sections:

1. Typical researcher development workflow
2. Typical lab developmental/production workflow
3. High Performance/High Throughput production workflow (local)
4. Improved HPC/HTC workflow with computing at a National Computing Facility.

2.1 Typical Researcher Development Workflow

The typical researcher development environment consists of a computer/workstation with a substantial CPU and Memory configuration attached to either an external storage device either directly via USB or on the network as a Network Attached Storage (NAS). The raw data which is obtained form the scanner as *DICOM* (*.dcm, *.dicom) slices is converted into *nifti* (*.nifti) file format and is stored on the storage devices attached to the researcher workstation. Due to variablilty in the quality of images received from different sources it is necessary to perform a Quality Control (QC) on the images before processing them and running them through the bootstrapped analysis. Since the bootstrapped analysis is performed on the *Fractional Anisotropy* (FA) map the QC is performed on the FA maps as well.

The process workflow is as described below and shown in Fig. 1.

a. Images are imported into an open source software TORTOISE (Tolerably Obsessive Registration and Tensor Optimization Indolent Software Ensemble) which is an open source software available from the National Institute of Heatlh (NIH).

   This import process makes the images available for further processing in TORTOISE.
b. Imported images are run through the DIFF_CALC process in TORTOISE (shown as Raw Calc in Fig. 1 since this process is run on the raw images). The DIFF_CALC not only provides the FA maps as output but also provides ROI based image noise estimate, and a number of other tensor derived quantities in analyze and nifti file formats.
c. Steps a and b are repeated until all the images are imported, converted and made available for QC.
d. A visual QC is performed on the FA maps by a team of a minimum of three researchers consisting of at least one physician and one engineer. The main aim of this QC is to look for imaging anomalies, distortions, artifacts, and the overall quality of the images. The images are scored individually on a scale of 1–3 with

Fig. 1 Typical researcher development workflow

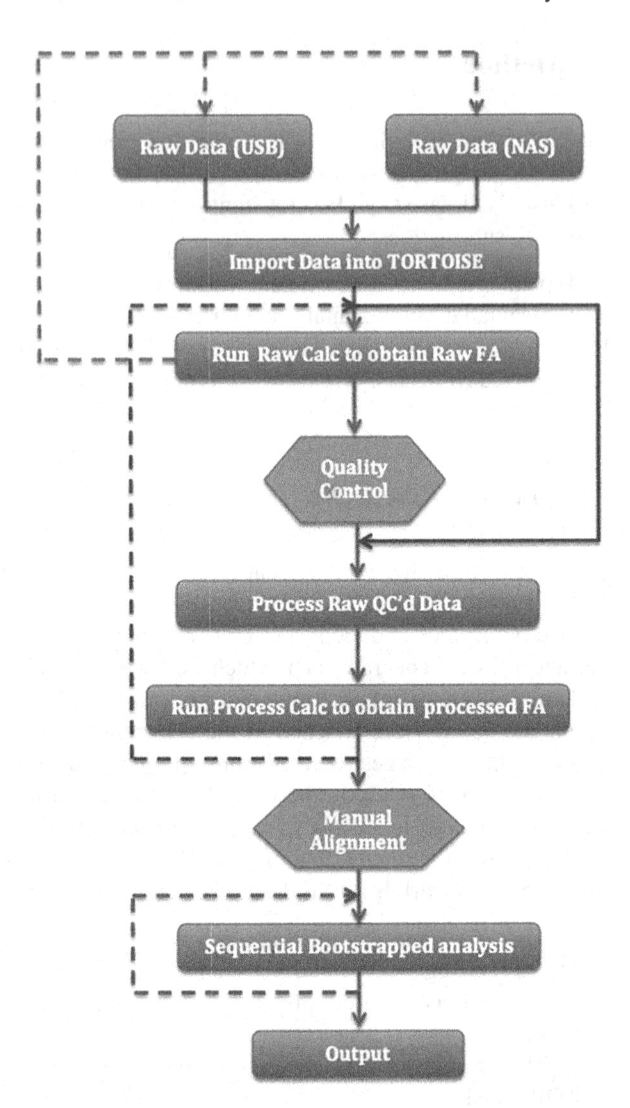

1 being an acceptable image and 3 being an unacceptable image. The average of all the scores are computed and only images with a final score less than 1.5 are chosen for further processing.

e. The imported images that have passed QC are then processed using the DIFF_PREP section in TORTOISE. The DIFF_PREP software computes the B-matrix of the image from gradient tables, compensates for motion and eddy current distortion with B-matrix reorientations, corrects B0 susceptibility indices EPI distortion, and reorients the images into a target space with B-matrix reorientation.

f. Processed images are run through the DIFF_CALC process in TORTOISE (shown as Process Calc in Fig. 1 since this process is run on the processed images). This gives out the processed FA maps for further analysis.
g. Steps e and f are repeated for all image that have passed QC to get the entire set of processed FA maps.
h. The processed FA maps are put through a manual alignment process to visually check the alignment of the images and apply correction. This is performed by another open source software AFNI developed by the NIH.
i. The processed and aligned FA maps are then run through further statistical process's such as a bootstrapped analysis and the output is obtained. The bootstrapped analysis is sequential and depends on the number of iterations required for bootstrapping.

2.2 Typical Lab Development/Production Workflow

The typical lab development/production environment consists of two or more computer workstations with a substantial CPU and Memory configuration attached to a Network Attached Storage (NAS) device. The raw data which is obtained form the scanner as *DICOM* (*.dcm, *.dicom) slices is converted into *nifti* (*.nifti) file format and is stored on the storage devices available from the lab workstations.

The process workflow is similar to the researchers development workflow obtained higher throughput by using larger number of machine for performing the higher compute cost processes such a the DIFF_CALC (step *b* and *f* above) and the DIFF_PREP (step *e* above). The typical lab workflow is as shown in Fig. 2. This method requires very little customization between the researchers development workflow and the production workflow but has overhead costs with respect to hardware purchases and network bandwidth availability for the NAS devices.

2.3 High Performance/High Throughput (HP/HT) Production Workflow (Local)

The typical HP/HT environment consists of a large number of compute nodes with a substantial CPU and Memory configuration attached to a High Performance storage devices. The raw data which is obtained form the scanner as *DICOM* (*.dcm, *.dicom) slices is converted into *nifti* (*.nifti) file format and is stored on the storage devices available from the lab compute nodes.

The above development and lab workflow can be massively parallelized and run on a High Performance/High Throughput compute cluster using the same application with little configuration. All application used in the development workflow can work on a compute cluster as well. Multiple VNC sessions were started on

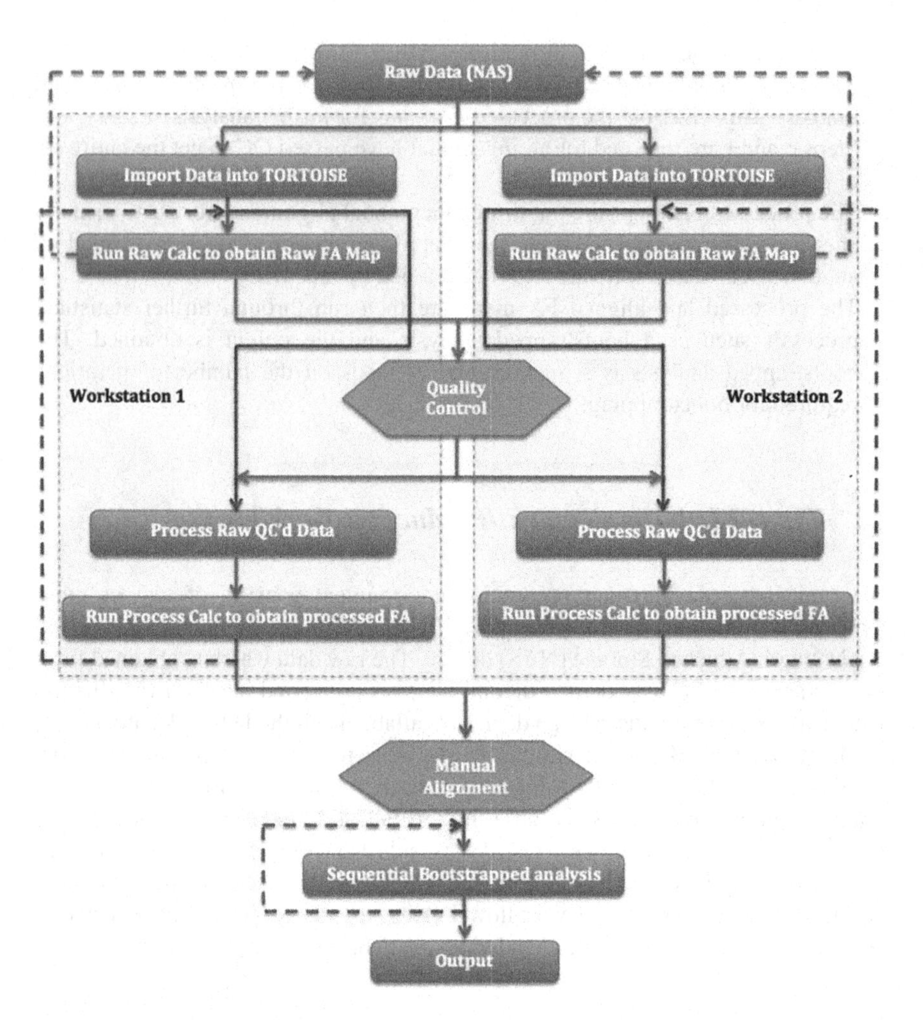

Fig. 2 Typical lab development/production workflow

the compute cluster in order to run GUI based applications such a DIFF_PREP and DIFF_CALC. They were distributed manually onto different compute nodes as *qlogin* jobs on a Sun Grid Engine (SGE) scheduler or an *sinteractive* job on a SLURM scheduler and a large number could be run in parallel as interactive jobs. The massively parallel workflow on the compute cluster was immensely helpful in reducing the time to QC and the time to bootstrapped analysis as will be shown in the results. The HP/HT production workflow is as shown in Fig. 3. The Same configuration was setup on UAB's new HPC cluster with the SLURM scheduler.

The major configuration and coding issue comes in the distributing of the bootstrapped analysis to fully utilize the power of High Performance computing. This problem was overcome by recoding the algorithm to be distributed and run in parallel for maximum efficiency.

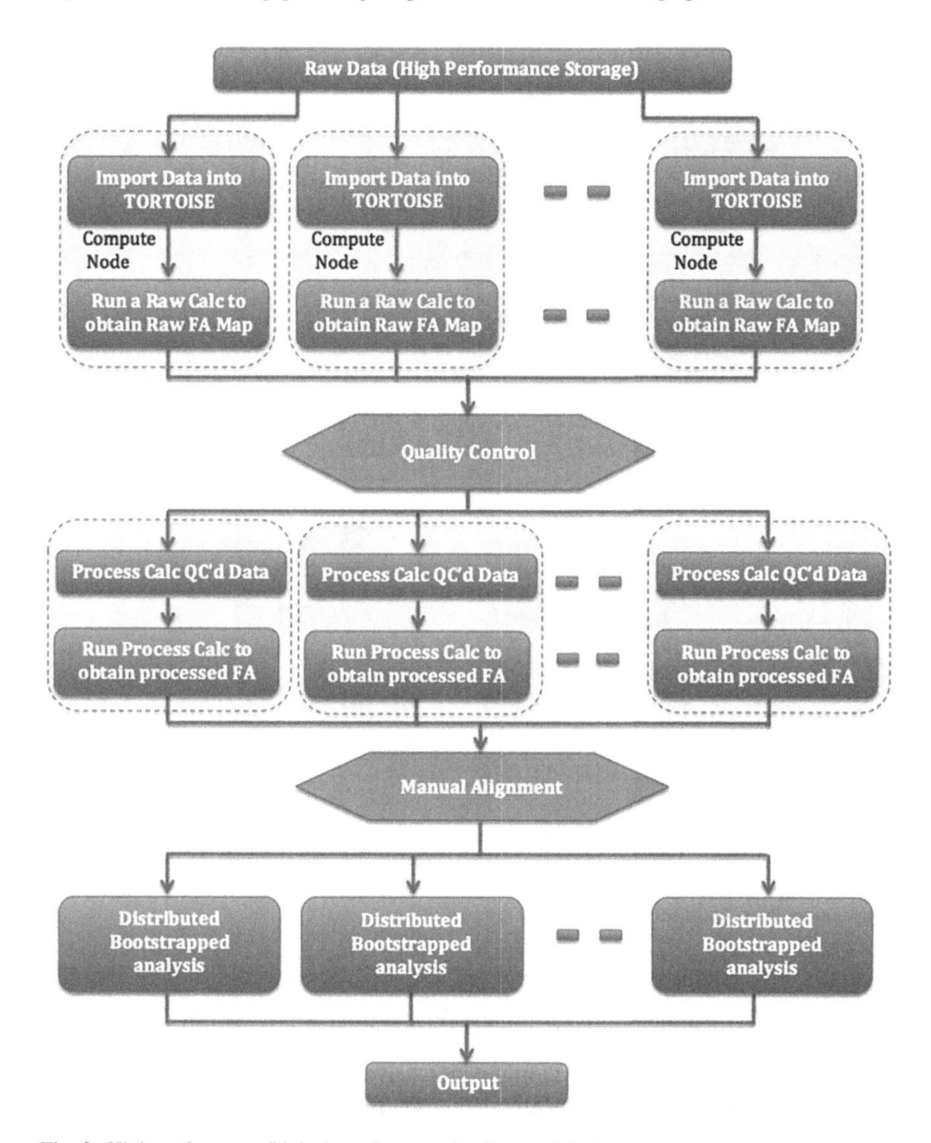

Fig. 3 High performance/high throughput production workflow

2.4 *Improved HPC/HTC Workflow with Computing Support from a National Computing Facility*

The process was distributed between the High Performance Computing fabrics at UAB and NICS through an XSEDE allocation for compute. The workflow is highly parallelized and involves large amount of data processing and movement. The massively parallel workflow on the compute cluster was immensely helpful in reducing

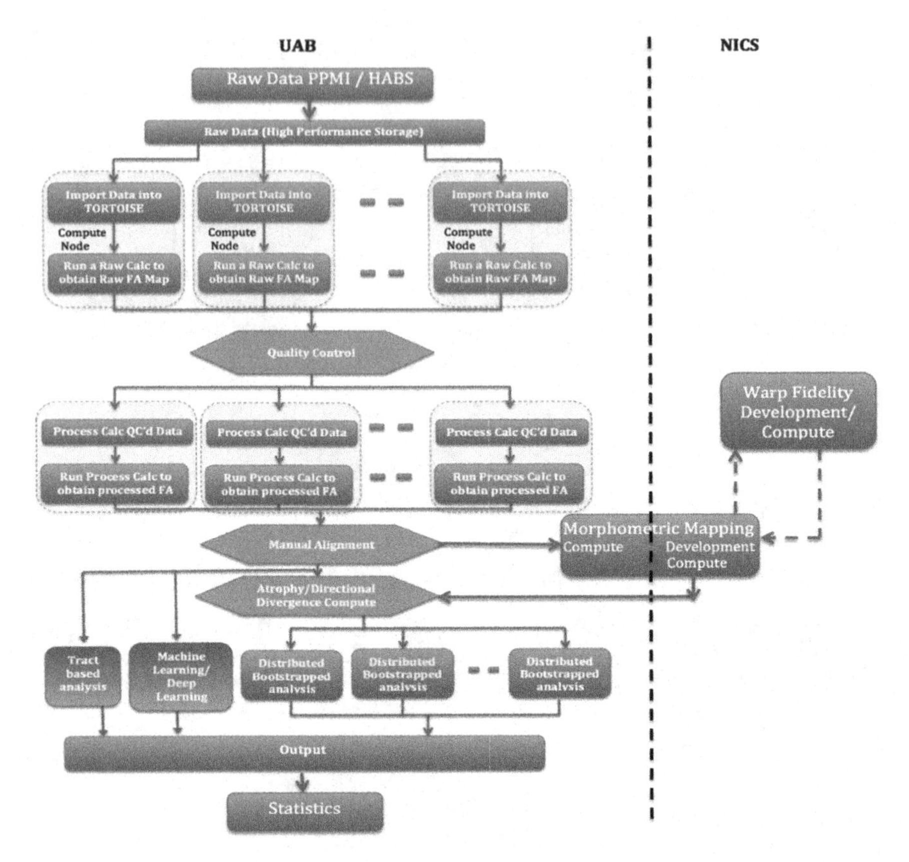

Fig. 4 High performance/high throughput production workflow with morphometric mapping and computation and support form a national computing center

the time to QC and the time to bootstrapped analysis as will be shown in the results. The HP/HT production workflow with support from a National Supercomputing Center helps the research add other techniques such as morphometric mapping and machine learning which are very compute intensive into the workflow as shown in Fig. 4. After the QC step a manual alignment is performed before sending the QC'd and aligned data (400 GB) to NICS for morphometric mapping.

The morphometric mapping is performed using a specialized version of the AFNI 3dQwarp code which has been improved to run efficiently in parallel down to a patch size of $3 \times 5 \times 3$ voxels. The default code is designed to map using a minimum patch size of $9 \times 9 \times 9$ voxels. This process is currently the most compute intensive step with the mapping down to $3 \times 5 \times 3$ voxels taking 300 compute-core hours per image.

The morphometric-mapped data (2 TB) is transferred back from the HPC system at NICS back to UAB for bootstrapping, machine learning, directional divergence atrophy calculation and statistical analysis. The output of the bootstrapping step (9 TB) is then statistically analyzed to produce results.

3 Test Setup, Assumptions and Results

The following hardware configurations were used for timing the workflow:

a. Typical researcher development workflow:
   Hardware 1 (HW1): one computer workstation—four cores (2.0 GHz Intel Xeon), 24 GB RAM, storage- 1 TB internal, 2 TB external HDD.
   Hardware 2 (HW2): one computer workstation—eight cores (2.93 GHz Intel i7), 16 GB RAM, storage- 2 TB internal, 2 TB external HDD.
b. Typical lab developmental/production workflow
   Hardware (HW1x2): two computer workstation—4 cores (2.0 GHz Intel Xeon), 24 GB RAM, storage- 1 TB internal, 2 TB external HDD. (each)
c. High Performance/High Throughput (HP/HT) production workflow:
   Hardware: UAB Cheaha HPC cluster—Using only Gen 2 nodes (sipsey) 48 nodes—12 cores (2.66 GHz Intel), 48–96 GB RAM, (each node) storage—500GB internal (each node) + 180 TB High Performance Lustre file system (available to all nodes)

3.1 Test Parameters

The tests for the same set of 100 subjects were run on first three workflows. Workflow iv has some additional processes that were incorporated such as the morphometric mapping step which was timed and are available as presented and published at XSEDE 2016 [16]. The results section deals with the time improvements until the preprocessing stage only.

The HP/HT timing results are shown using a single node to show performance improvement over a single Lab system and 10 nodes on the HPC system to show overall improvement in preprocessing throughput.

One working day is considered to be 8 h and processes running under 24 h are considered to be completed in one day.

All the time shown are average times and rounded off to the closest integer.

Accept rate after Quality Control was close to 2/3 and the timing for further steps was using 66% as number of subjects whose image quality was deemed acceptable for further processing.

Overhead is assumed to be between 10–15 min for manually setting up the parallel processing streams

4 Discussion and Conclusion

The results as shown in Table 1 show that preprocessing MRI images to for use in a Neuroscience workflow can be brought down from months to under two weeks improving the ability of researchers to perform further analysis quickly or

Table 1 Preprocessing workflow timing on different hardware profiles

Step	Hardware 1 (serial)	Hardware 2 (serial)	Lab (HW 1 × 2 parallel)	HP/HT (single node)	HP/HT (10 nodes parallel)
Import single (avg time/subject—minutes)	10	8	10	4	NA
Import 100 subjects	1000 (~16 h)	800 (~13 h)	500 (~8 h)	400 (~7 h)	40 + overhead (~1 h)
Raw Calc (avg time/subject—minutes)	19	14	19	15	NA
Raw Calc 100 subjects—minutes	1900 (~32 h)	1400 (~23 h)	950 (~16 h)	1500 (~25 h)	150 + overhead (~3 h)
A = Total Time to QC (100 subjects)	48 h (6 working days)	36 h (4.5 working days	24 h (3 working days)	32 h (4 working days)	4 h (0.5 working days)
Visual QC (100 subjects—minutes)	300 (~5 h)	300 (~5 h)	300 (~5 h)	300 (~5 h)	300 (~5 h)
Raw Process (avg time/subject—hours)	18 (one per day)	13 (one per day)	18 (two per day)	14 (one per day)	NA
B = Raw Process – 66 subject (66% accept rate - hours)	1188 (~66 working days)	858 (~66 working days)	1188 (~33 working days)	924 (~66 working days)	924 (~7 working days)
Process Calc (avg time/subject—minutes)	19	14	19	15	NA
C = Process Calc 66 subjects—minutes	1254 (~21 h) (~2.5 working days)	924 (~15 h) (~2 working days)	627 (~10 h) (~1 working day)	990 (~16 h) (~2 working days)	99 + overhead (~2 h) (~0.25 working days)
Total time for preprocessing 100 subjects (A + B + C)	75 working days ~15 weeks	73 working days ~14.5 weeks	37 working days ~7.5 weeks	72 working days ~14.5 weeks	8 working days ~1.5 weeks

integrate more complex processes in the workflow with out waiting for results to be available months later. In case of workflow changes or errors development time lost is also reduced. Researchers can integrate more data into their analysis pipelines and use data driven analytics to a much higher extent than can be done by traditional research methodologies. In the twenty-first century, we now can potentially analyze large information flows derived from individual patients, allowing physicians enter an era of personalized medicine. Personalized medicine has the potential of decreasing adverse events by increasing the amount of data and derived information on individual patients. Benefits of personalized medicine are all ready starting to become available. A key feature, however, in these analyses is that a large number of features derived from large amounts of data (such as genetic data) often must be analyzed in order to capture the interaction between complex diseases and the individual patients who suffer from these diseases.

Similar to genetic information, brain imaging provides large amounts of data with high dimensionality that must be managed in order to obtain useful, patient-specific information. We present a workflow study that shows the practical implications of using an HPC platform in a research setting to improve imaging analysis. Specifically, our analysis shows a tenfold improvement in turn-around of results in the limited setting of replacing a traditional one machine analysis tree with a ten-node computational platform. Note that if we add more data, we simply need to add more nodes to maintain an identical (in this case 1.5 week) turn around for any new analysis using this workflow.

In summary, we illustrate in a practical manner the value of utilizing a HPC platforms to improve a brain imaging research processes. We expect that advances in computational speed, and in computational methods (such as adding learning networks and improved statistical methods), will continue to improve reported efficiency gains. This peper support that transitioning research processes associated with MRI to HPC platforms can accelerate discovery science. Accelerating discovery science, in turn, may allow brain imaging to become part of analytics and biomarker profiles that will underpin a developing era of personalized medicine.

References

1. Rinck, P.: Magnetic resonance in medicine. In: The Basic Textbook of the European Magnetic Resonance Forum, 9th edn, (2016)
2. Tonellato, P.J., Crawford, J.M., Bogusky, M.S., Safitz, J.E.: A national agenda for the future of pathology in personalized medicine: report of the proceedings of a meeting at the Banbury conference center on genome-era pathology, precision diagnostics, and preemptive care: a stakeholder summit. Am. J. Clin. Pathol. **135**(5), 668–672 (2011)
3. Skidmore, F., Yang, M., von Deneen, K., Collingwood, J., He, G., White, K., Korenkevych, D., Savenkov, A., Heilman, K., Gold, M., Liu, Y.: Reliability analysis of the resting state can sensitively and specifically identify the presence of parkinson disease. NeuroImage. **75**, 249–261 (2013)
4. Ma, Y., Huang, C., Dyke, J., Pan, H., Alsop, D., Feigin, A., et al.: Parkinson's disease spatial covariance pattern: noninvasive quantification with perfusion mri. J. Cereb. Blood Flow Metab. **30**, 505–509 (2010)

5. Skidmore, F., Spetsieris, P., Anthony, T., Cutter, G., von Deneen, K., Liu, Y., White, K., Heilman, K., Myers, J., Standaert, D., Lahti, A., Eidelberg, D., Ulug, A.: A full-brain, bootstrapped analysis of diffusion tensor imaging robustly differentiates parkinson disease from healthy controls. Neuroinformatics. **13**, 7–18 (2015)
6. Gorell, J., Ordidge, R., Brown, G., Deniau, J., Buderer, N., Helpern, J.: Increased iron-related mri contrast in the substantia nigra in parkinson's disease. Neurology. **45**, 1138–1143 (1995)
7. Michaeli, S., Oz, G., Sorce, D., Garwood, M., Ugurbil, K., Majestic, S., Tuite, P.: Assessment of brain iron and neuronal integrity in patients with parkinson's disease using novel mri contrasts. Mov. Disord. **22**, 334–340 (2007)
8. Lazar, N.: The Statistical Analysis of Functional MRI Data. Springer-Verlag, New York (2008)
9. Eklund, A., et al.: Empirically investigating the statistical validity of SPM, FSL and AFNI for single subject fMRI analysis. In: IEEE 12th International Symposium on Biomedical Imaging (2015)
10. Barr, W., Morrison, C.: Handbook on the Neuropsychology of Epilepsy. Springer, New York (2010)
11. Laureys, S., Gosseries, O., Tononi, G.: The Neurology of Consciousness: Cognitive Neuroscience and Neuropathology. Academic Press, Cambridge, MA (2015)
12. AFNI package. http://afni.nimh.nih.gov/afni/
13. R. Cox AFNI: What a long strange trip it's been. NeuroImage. **62**, 743–747 (2012)
14. R. Cox AFNI: Software for analysis and visualization of functional magnetic resonance neuroimages. Comput. Biomed. Res. **29**, 162–173 (1996)
15. Pierpaoli, L., Walker, M., Irfanoglu, O., Barnett, A., Basser, P., Chang, L-C., Koay, C., Pajevic, S., Rohde, G., Sarlls, J., Wu, M.: TORTOISE: an integrated software package for processing of diffusion MRI data. In: ISMRM 18th Annual Meeting, Stockholm, Sweden, #1597 (2010)
16. Yin, J., Anthony, T., Marstrander, J., Liu, Y., Burdyshaw, C., Horton, M., Crosby, L.D., Brook, R.G., Skidmore, F.: Optimization of non-linear image registration in AFNI. In: Proceedings of the XSEDE16 Conference on Diversity, Big Data, and Science at Scale (p. 6). ACM (2016)

Index

S.C. Suh, T. Anthony (eds.), *Big Data and Visual Analytics*,
https://doi.org/10.1007/978-3-319-63917-8

Printed in the United States
By Bookmasters